RIBA Plan of
Information E

**The RIBA Plan of Work 2013 Guides**

Other titles in the series:

*Design Management*, by Dale Sinclair
*Contract Administration*, by Ian Davies
*Project Leadership*, by Nick Willars
*Town Planning*, by Ruth Reed

Coming soon:

*Sustainability*, by Sandy Halliday and Richard Atkins
*Conservation*, by Hugh Feilden
*Health and Safety*, by Peter Caplehorn

The RIBA Plan of Work 2013 is endorsed by the following organisations:

| Royal Incorporation of Architects in Scotland | Chartered Institute of Architectural Technologists | Royal Society of Architects in Wales | Construction Industry Council | Royal Society of Ulster Architects |

RIBA Plan of Work 2013 Guide

# Information Exchanges

**Richard Fairhead**

RIBA ⫞ **Publishing**

© RIBA Enterprises Ltd, 2015
Published by RIBA Publishing, The Old Post Office, St Nicholas Street,
Newcastle upon Tyne NE1 1RH

ISBN 978 1 85946 554 7
Stock code 82564

The right of Richard Fairhead to be identified as the Author of this
Work has been asserted in accordance with the Copyright, Designs
and Patents Act 1988 sections 77 and 78.

British Library Cataloguing in Publication Data
A catalogue record for this book is available from the British Library.

Commissioning Editor: Sarah Busby
Series Editor: Dale Sinclair
Project Manager: Alasdair Deas
Design: Kneath Associates
Typesetting: Academic+Technical, Bristol, UK
Printed and bound by CPI Group (UK) Ltd
Cover image: © Neil Young/Foster + Partners

While every effort has been made to check the accuracy and quality
of the information given in this publication, neither the Author nor the
Publisher accept any responsibility for the subsequent use of this
information, for any errors or omissions that it may contain, or for any
misunderstandings arising from it.

RIBA Publishing is part of RIBA Enterprises Ltd
www.ribaenterprises.com

# Contents

# Foreword

In 2015 the construction industry is at a crossroads, in part influenced by the 2011 Government mandate for 'Level 2' BIM by 2016, and in part by the need to embrace emerging technologies to support building design, construction and operation. The drive towards BIM as standard working practice by 2016 is an achievable goal for all but it requires largely unconnected groups to work more closely to deliver this common aim. Something not often embraced within our industry.

Since the Government BIM mandate was launched significant work has been undertaken by a range of bodies to define a set of achievable standards with which to provide commonality and foster greater collaboration across the building industry. This has tested existing working practices and terminology, highlighting inconsistencies or conflict between many well-established standards and processes. What has emerged is a set of unified standards and processes that more fully support UK BIM implementation.

One of the core documents that has emerged since 2011 is the RIBA Plan of Work 2013. Richard Fairhead and I first met when we sat on the Plan of Work review panel. Our mutual interest in the process of designing and delivering buildings coupled to emerging BIM technologies was evident.

It is clear that BIM is a key part of the necessary evolution of the architectural profession and the wider construction industry. What must not be lost is that BIM only plays a part in the design, delivery and operation of buildings. In my opinion the current emphasis it is given is detracting from the overall quality of the built form.

Richard has set the suite of UK Level 2 BIM standards in the context of the overall design process, as delivery tools rather than an outcome in themselves. This is grounded by the RIBA Plan of Work 2013 and provides, for the first time, an understandable and comprehensive manual to support building design and delivery. I am clear that this, along with its companion guides, will serve as a standard text moving forward, allowing architects to confidently adopt the appropriate new standards and technologies demanded by Level 2 BIM.


**Alistair Kell**
*Director of Information and Technology at BDP*

# Series editor's foreword

The RIBA Plan of Work 2013 was developed in response to the needs of an industry adjusting to emerging digital design processes, disruptive technologies and new procurement models, as well as other drivers. A core challenge is to communicate the thinking behind the new RIBA Plan in greater detail. This process is made more complex because the RIBA Plan of Work has existed for 50 years and is embodied within the psyche and working practices of everyone involved in the built environment sector. Its simplicity has allowed it to be interpreted and used in many ways, underpinning the need to explain the content of the Plan's first significant edit. By relating the Plan to a number of commonly encountered topics, the *RIBA Plan of Work 2013 Guides* series forms a core element of the communication strategy and I am delighted to be acting as the series editor.

The first strategic shift in the RIBA Plan of Work 2013 was to acknowledge a change from the tasks of the design team to those of the project team: the client, design team and contractor. Stages 0 and 7 are part of this shift, acknowledging that buildings are used by clients, or their clients, and, more importantly, recognising the paradigm shift from designing for construction towards the use of high-quality design information to help facilitate better whole-life outcomes.

New procurement strategies focused around assembling the right project team are the beginnings of significant adjustments in the way that buildings will be briefed, designed, constructed, operated and used. Design teams are harnessing new digital design technologies (commonly bundled under the BIM wrapper), linking geometric information to new engineering analysis software to create a generation of buildings that would not previously have been possible. At the same time, coordination processes and environmental credentials are being improved. A core focus is the progressive fixity of high-quality information – for the first time, the right information at the right time, clearly defining who does what, when.

The RIBA Plan of Work 2013 aims to raise the knowledge bar on many subjects, including sustainability, Information Exchanges and health and safety. The *RIBA Plan of Work 2013 Guides* are crucial tools in disseminating and explaining how these themes are fully addressed and how the new Plan can be harnessed to achieve the new goals and objectives of our clients.

**Dale Sinclair**
*June 2015*

# Acknowledgements and dedication

My thanks are extended to Matthew Scammels, PJ O'Sullivan, Phil Oakes, Paul Green and Nigel Ostime for allowing me to bend their ear occasionally.

I would also like to thank Dale Sinclair for chairing and driving the BIM overlay and Plan of Work 2013 groups and for providing the opportunity to contribute to this series. I would also like to thank Dale and Sarah Busby for their continued help and support throughout the development of this guide and all the others who assisted in its production.

Finally, I would like thank my wife, Lisa, for all her support and time doing the things I didn't, and my children Elliot, Callum and Lia for their continued patience with me not seeing them score tries, ride horses or decorate their rooms.

## Picture credits

The following items are reproduced with permission. Figure 0.1: *Bew/ Richards Maturity Diagram*, Mervyn Richards; 1.2: US Army Engineer Research and Development Center (ERDC) Public Affairs Office; 1.3, 1.4, 1.8, 7.2: The NBS; 1.5: Construction Industry Council, *Building Information Model (BIM) Protocol* (2013); 2.5, 3.2: BSI, BS 1192: 2007; 3.1: Mervyn Richards, *Building Information Management. A Standard Framework and Guide to BS 1192*, British Standards Institution (2010); 3.4, 3.6: BSI, PAS 1192-2: 2013; 7.1: BSI, PAS 1192-3: 2014; 3.3: AEC Committee, *AEC (UK) CAD standard for drawing management* (2005); 3.8: AEC Committee, *AEC (UK) BIM Protocol* (2012): 4.1: Health and Safety Executive, *Integrated gateways: planning out health & safety risk* (2004); Table 0.1: Cabe/ Design Council, *The Value Handbook* (2006).

Permission to reproduce extracts from British Standards is granted by BSI Standards Limited (BSI). No other use of this material is permitted. British Standards can be obtained in PDF or hard copy formats from the BSI online shop: http://www.bsigroup.com/Shop

# About the author

Richard is a qualified architect with extensive knowledge and experience of designing and delivering within a wide range of sectors, on both large and small projects, including the implementation of information management systems. Richard has been keenly involved in the development of design through the use of computers and CAD for the past 20 years. He has developed a number of IT CAD protocols and execution plans, as well as developing the use of BIM at a number of practices.

Richard was a key member of the RIBA working group that developed the new RIBA Plan of Work 2013, having already been involved in the *BIM Overlay to the RIBA Outline Plan of Work* in 2012. He has presented widely on the development of BIM at several RIBA, RIUA and Bentley Systems events, and has previously written on BIM in the RIBA's *Handbook of Practice Management*.

# About the series editor

Dale Sinclair is Director of Technical Practice for AECOM's architecture team in EMEA. He is an architect and was previously a director at Dyer and an associate director at BDP. He has taught at Aberdeen University and the Mackintosh School of Architecture and regularly lectures on BIM, design management and the RIBA Plan of Work 2013. He is passionate about developing new design processes that can harness digital technologies, manage the iterative design process and improve design outcomes.

He is currently the RIBA Vice President, Practice and Profession, a trustee of the RIBA Board, a UK board member of BuildingSMART and a member of various CIC working groups. He was the editor of the *BIM Overlay to the Outline Plan of Work 2007*, edited the RIBA Plan of Work 2013 and was author of its supporting tools and guidance publications: *Guide to Using the RIBA Plan of Work 2013* and *Assembling a Collaborative Project Team*.

# Introduction

## Overview

The exponential expansion of the internet has enabled vast amounts of data to be created and utilised every second. Much of it will rarely be used more than once. However, its presence represents a significant shift in the way we want to understand and access information and, perhaps more importantly, the speed at which we expect it to be available.

The construction industry, reacting to these digital developments, has adapted its core processes to align with these emerging technologies, changing the way that project information is created, managed and used. This shift, to predominantly digital systems, will put information at the core of everything we do, from the initial concept to the finished product in use.

We are currently 'in between' two distinct processes, with project information being produced using either CAD-based systems or emerging information-centric processes, such as BIM (Building Information Modelling), or, in this transitional period, a mixture of the two.

The RIBA Plan of Work was previously structured around tasks that underpinned the design and delivery process. However, as production and management systems have evolved, the 2013 Plan now places a specific focus on the requirements and content of Project Information, outlining what should be produced and, more specifically, when it should be produced, in support of a collaborative approach. These Information Exchanges define the purpose of the Project Information – how it can be used and what it should be used for – in order to develop the design solution and operate and maintain the facility.

In summary these Information Exchanges are outlined as:

Stage 0  Strategic Brief
Stage 1  Initial Project Brief
Stage 2  Final Project Brief, Concept Design (including outline structural and buildings services design), Project Strategies and Cost Information
Stage 3  Developed Design (including the coordinated architectural, structural and building services design) and updated Cost Information
Stage 4  Technical Design
Stage 5  'As-constructed' Information
Stage 6  updated 'As-constructed' Information
Stage 7  Feedback and asset information.

## Context

At the outset, information will typically be used to define the client's needs and aspirations and to develop the overall approach. The briefing process will identify these early requirements using Feedback, research, Site Information and benchmarking, as well as framing the developing design, project team and scope throughout Stages 0, 1 and 2. The Project Execution Plan will outline the key Project Strategies in support of each Information Exchange, including protocols, processes and best practice standards for structuring and managing the Project Information.

Through the design stages – Stages 2, 3 and 4 – a managed process can accurately progress information collaboratively within the design team, assist coordination, improve production efficiencies and enhance both performance and quality. As the information matures it will be critical for the design team to understand what data is valid and appropriate, who develops it and, more crucially, who is responsible for the key interfaces. This is of particular importance when considering the integration of information produced by specialist designers and subcontractors and how the status and development of the design will be affected.

The completion of the project raises similar responsibility issues, with different procurement methods, roles and scopes requiring different approaches to the production and validation of the 'As-constructed' Information. How this is managed, both in terms of appointment and the handover of information following Stages 5 and 6, will be critical to the success of the project in use.

The handover and occupation of the project will be based on the information collected and collated in Stages 5, 6 and 7, validating the performance of the systems in use and the success of the original design criteria. This information can also be utilised to assess whether the Project Outcomes have been achieved.

Post-occupancy Evaluation and building performance evaluation can provide a unique understanding of how operational data and information can be used within similar projects as well as gauging the actual performance of installed systems throughout the lifespan of the project. These processes ensure that key learning points can be incorporated into Feedback for the client and for the industry as a whole, underpinning the importance of connecting Stages 0 and 7 and premised on positive outcomes, reinforcing the RIBA Plan of Work's cyclical approach to the management and use of information.

## How to use this book

Properly defined Information Exchanges ensure that the right information is produced at the right time and so are a fundamental starting point for defining the overall scope of the project and the project team. They represent a formal issue of information for review and sign-off by the client at each stage of the project, but can be supplemented by any number of informal Information Exchanges between the project team members throughout the design, construction and in-use processes.

The structure of this guide follows the RIBA Plan of Work 2013 stages from 0 to 7, with each chapter exploring how the Core

Objectives, Key Support Tasks and associated Project Strategies impact the overall information requirements of the project as it develops.

The guide focuses primarily on the Information Exchanges task bar, setting out the completion requirements for each stage, as well as what information might typically be exchanged to collaboratively develop, construct and operate the project. The Information Exchanges also acknowledge that each project will be different and that the information requirements will vary based on the project specifics, the constraints and the client's objectives.

The structure and format of the Project Information and the way it is managed throughout the project is as important as the final output. Irrespective of the systems and software used to progress the design, it is essential to ensure that the standards and protocols used to produce and manage the Project Information are clearly defined from the outset. These procedures will enable the format of the information to be understood, validated and used appropriately, avoiding unnecessary waste and any conflicts and inconsistencies that could lead to significant issues on site.

Understanding how information is created, exchanged, managed and delivered has resulted in considerable reductions in design and construction programmes, operational costs and waste. More importantly, these processes allow the project team to develop and manage the project collaboratively, leading to better Project Outcomes and a higher quality of delivered design.

Richard Fairhead
*June 2015*

## Using this series

For ease of reference each book in this series is broken down into chapters that map on to the stages of the Plan of Work. So, for instance, the first chapter covers the tasks and considerations around Information Exchanges at Stage 0.

We have also included several in-text features to enhance your understanding of the topic. The following key will explain what each icon means and why each feature is useful to you:

 The 'Example' feature explores an example from practice, either real or theoretical

 The 'Tools and Templates' feature outlines standard tools, letters and forms and how to use them in practice

 The 'Signpost' feature introduces you to further sources of trusted information from books, websites and regulations

 The 'Definition' feature explains key terms in this topic area in more detail

 The 'Hints and Tips' feature dispenses pragmatic advice and highlights common problems and solutions

 The 'Small Project Observation' feature highlights useful variations in approach and outcome for smaller projects

# RIBA ⚜

The **RIBA Plan of Work 2013** organises the process of briefing, designing, constructing, maintaining, operating and using building projects into a number of key stages. The content of stages may vary or overlap to suit specific project requirements.

**RIBA Plan of Work 2013**

| Tasks ▼ | Stages ▶ 0 **Strategic Definition** | 1 **Preparation and Brief** | 2 **Concept Design** | 3 **Developed Design** |
|---|---|---|---|---|
| Core Objectives | Identify client's **Business Case** and **Strategic Brief** and other core project requirements. | Develop **Project Objectives**, including **Quality Objectives** and **Project Outcomes**, **Sustainability Aspirations**, **Project Budget**, other parameters or constraints and develop **Initial Project Brief**. Undertake **Feasibility Studies** and review of **Site Information**. | Prepare **Concept Design**, including outline proposals for structural design, building services systems, outline specifications and preliminary **Cost Information** along with relevant **Project Strategies** in accordance with **Design Programme**. Agree alterations to brief and issue **Final Project Brief**. | Prepare **Developed Design**, including coordinated and updated proposals for structural design, building services systems, outline specifications, **Cost Information** and **Project Strategies** in accordance with **Design Programme**. |
| Procurement *Variable task bar | Initial considerations for assembling the project team. | Prepare **Project Roles Table** and **Contractual Tree** and continue assembling the project team. | ◀— The procurement strategy does not fundamentally alter the progression of the design or the level of detail prepared at | a given stage. However, **Information Exchanges** will vary depending on the selected procurement route and **Building Contract**. A bespoke **RIBA Plan of Work** |
| Programme *Variable task bar | Establish **Project Programme**. | Review **Project Programme**. | Review **Project Programme**. | ◀— The procurement route may dictate the **Project Programme** and result in certain stages overlapping |
| (Town) Planning *Variable task bar | Pre-application discussions. | Pre-application discussions. | ◀— Planning applications are typically made using the Stage 3 output. | A bespoke **RIBA Plan of Work 2013** will identify when the |
| Suggested Key Support Tasks | Review **Feedback** from previous projects. | Prepare **Handover Strategy** and **Risk Assessments**. Agree **Schedule of Services**, **Design Responsibility Matrix** and **Information Exchanges** and prepare **Project Execution Plan** including **Technology** and **Communication Strategies** and consideration of **Common Standards** to be used. | Prepare **Sustainability Strategy, Maintenance and Operational Strategy** and review **Handover Strategy** and **Risk Assessments**. Undertake third party consultations as required and any **Research and Development** aspects. Review and update **Project Execution Plan**. Consider **Construction Strategy**, including offsite fabrication, and develop **Health and Safety Strategy**. | Review and update **Sustainability, Maintenance and Operational** and **Handover Strategies** and **Risk Assessments**. Undertake third party consultations as required and conclude **Research and Development** aspects. Review and update **Project Execution Plan**, including **Change Control Procedures**. Review and update **Construction** and **Health and Safety Strategies**. |
| Sustainability Checkpoints | **Sustainability Checkpoint — 0** | **Sustainability Checkpoint — 1** | **Sustainability Checkpoint — 2** | **Sustainability Checkpoint — 3** |
| Information Exchanges (at stage completion) | **Strategic Brief.** | **Initial Project Brief.** | **Concept Design** including outline structural and building services design, associated **Project Strategies**, preliminary **Cost Information** and **Final Project Brief**. | **Developed Design**, including the coordinated architectural, structural and building services design and updated **Cost Information**. |
| UK Government Information Exchanges | Not required. | Required. | Required. | Required. |

*Variable task bar – in creating a bespoke project or practice specific RIBA Plan of Work 2013 via www.ribaplanofwork.com a specific bar is selected from a number of options.

The **RIBA Plan of Work 2013** should be used solely as guidance for the preparation of detailed professional services contracts and building contracts.

**www.ribaplanofwork.com**

| 4 | 5 | 6 | 7 |
|---|---|---|---|
| **Technical Design** | **Construction** | **Handover and Close Out** | **In Use** |
| Prepare **Technical Design** in accordance with **Design Responsibility Matrix** and **Project Strategies** to include all architectural, structural and building services information, specialist subcontractor design and specifications, in accordance with **Design Programme**. | Offsite manufacturing and onsite **Construction** in accordance with **Construction Programme** and resolution of **Design Queries** from site as they arise. | Handover of building and conclusion of **Building Contract**. | Undertake **In Use** services in accordance with **Schedule of Services**. |
| **2013** will set out the specific tendering and procurement activities that will occur at each stage in relation to the chosen procurement route. | Administration of **Building Contract**, including regular site inspections and review of progress. | Conclude administration of **Building Contract**. | |
| or being undertaken concurrently. A bespoke **RIBA Plan of Work 2013** will clarify the stage overlaps. | The **Project Programme** will set out the specific stage dates and detailed programme durations. | | |
| planning application is to be made. | | | |
| Review and update **Sustainability, Maintenance and Operational** and **Handover Strategies** and **Risk Assessments**.<br><br>Prepare and submit Building Regulations submission and any other third party submissions requiring consent.<br><br>Review and update **Project Execution Plan**.<br><br>Review **Construction Strategy**, including sequencing, and update **Health and Safety Strategy**. | Review and update **Sustainability Strategy** and implement **Handover Strategy**, including agreement of information required for commissioning, training, handover, asset management, future monitoring and maintenance and ongoing compilation of **'As-constructed' Information**.<br><br>Update **Construction** and **Health and Safety Strategies**. | Carry out activities listed in **Handover Strategy** including **Feedback** for use during the future life of the building or on future projects.<br><br>Updating of **Project Information** as required. | Conclude activities listed in **Handover Strategy** including **Post-occupancy Evaluation**, review of **Project Performance, Project Outcomes** and **Research and Development** aspects.<br><br>Updating of **Project Information**, as required, in response to ongoing client **Feedback** until the end of the building's life. |
| **Sustainability Checkpoint — 4** | **Sustainability Checkpoint — 5** | **Sustainability Checkpoint — 6** | **Sustainability Checkpoint — 7** |
| Completed **Technical Design** of the project. | **'As-constructed' Information**. | Updated **'As-constructed' Information**. | **'As-constructed' Information** updated in response to ongoing client **Feedback** and maintenance or operational developments. |
| Not required. | Not required. | Required. | As required. |

© RIBA

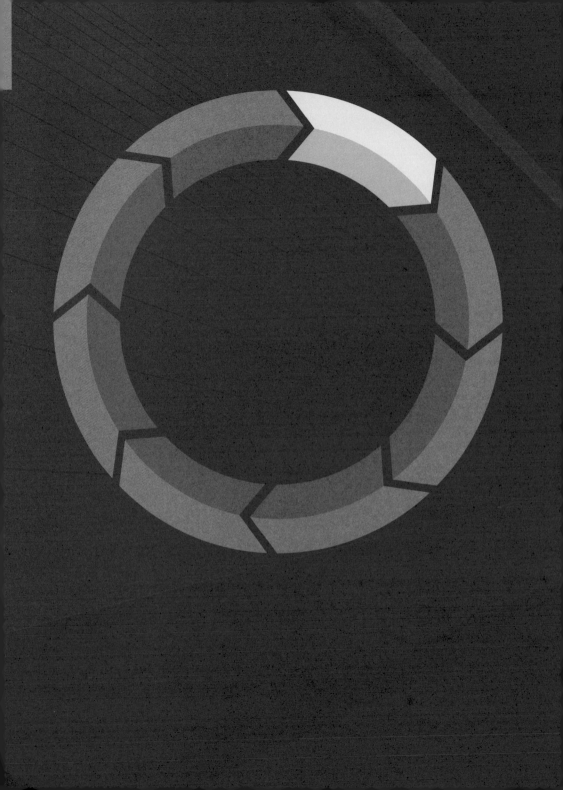

Stage 0

# Strategic
# Definition

# Chapter overview

During this initial stage the tasks and information necessary to determine the feasibility, aspirations and outcomes of the project will be collated and reviewed. Assessing Feedback from similar projects, best practice research, Post-occupancy Evaluations and performance of systems in use will outline the principles required to develop and finalise a strategic appraisal. The Strategic Brief will record this approach and provide a solid foundation from which the development of the project, if appropriate, can be progressed.

**The key coverage in this chapter is as follows:**

Developing a strategic approach

The Business Case

Exploring Project Outcomes and requirements

Documenting the Strategic Brief

What are the Information Exchanges at the end of Stage 0?

# Introduction

Stage 0 is a new stage, during which the viability of a project is assessed, typically before a formal response is developed and large amounts of time and effort have been expended.

The project team will focus on the development of the Strategic Brief; an 'ends-driven' document that succinctly sets out the client's key outcomes and achievable objectives, including:

what the client requires

why it is important

how it might be achieved.

The Strategic Brief will not necessarily be a large document, but it will establish the appropriate information required to:

understand whether a project is needed

select and scope the project team

set out the strategic objectives

be approved, signed off and 'owned' by the client.

The final document will respond to the client's Business Case, outlining a robust understanding of the key processes and requirements.

## What are the Core Objectives of this stage?

The Core Objectives of the RIBA Plan of Work 2013 at Stage 0 are:

| Tasks ▼ | **0** Strategic Definition |
|---|---|
| Core Objectives | Identify client's **Business Case** and **Strategic Brief** and other core project requirements. |

At this stage, the Core Objectives are structured solely in support of the Strategic Brief, with much of the information developed from client-specific requirements, although other studies, such as precedents, research, performance data and Feedback, should be reviewed to test the Business Case, especially if the client is new to the design process. This approach will ensure that the Initial Project Brief can be developed appropriately within Stage 1 alongside any Feasibility Studies and prior to the Concept Design.

## Developing a strategic approach

Some of the most important project decisions are made during this stage, prior to the commencement of any design work, and it is highly likely that much of the success and value of the project will be determined during these early explorations. For example, the strategic appraisal may highlight that there is no need for a project at all, with the client's requirements being achievable through an alternative route, or that a refurbishment approach is appreciably more viable and appropriate than a new-build facility. These early decisions are significant in defining the project direction and will help to minimise effort wasted on developing inappropriate briefs and concept designs.

Given that these early decisions will have a fundamental impact on subsequent developments, a comprehensive evaluation of the project's potential needs to be completed in order to demonstrate where value can be added to a client's operations and aspirations.

This strategic appraisal will need to address:

I the context and constraints of probable sites
I specific building sector best practice (eg commercial, health, residential, retail)
I benchmarking data from similar projects
I the nature and extent of the project
I the Business Case
I how the project team will be defined and what services will be necessary throughout the subsequent stages
I the implications and benefits of using a Building Information Model (BIM) model to design and deliver the project.

A robust understanding of these issues will enable the Initial Project Brief to be developed appropriately at the next stage, which will provide the basis from which any Feasibility Studies and the Concept Design can be developed.

## The Business Case

The Business Case for a project is the rationale behind the initiation of a new building project. It may consist solely of a reasoned argument. It may contain supporting information, financial appraisals or other background information. It should also highlight initial considerations for the Project Outcomes.

The Business Case might contain:

I the appraisals from a number of sites
I the drivers behind the proposed project and the outcomes that are being sought
I a high-level cost study
I examples of suitable precedents, to allow quality aspirations to be understood.

The contents will vary from project to project; however, it is crucial that any drivers or influencing factors are recorded and signed off by the client.

In summary, it is a combination of objective and subjective considerations. The Business Case might be prepared in relation to, for example, appraising a number of sites or in relation to assessing a refurbishment against a new-build option.

## Exploring Project Outcomes and requirements

The Strategic Definition stage should be considered as 'beginning with the end in mind'. It is inherently dependent on the information and outputs collected from similar projects while they are in use. Collating this data can equally be considered as 'ending with the beginning in mind'. This cyclical approach to understanding Project Outcomes and Feedback is essential to setting out the aspirations and performance requirements within the Strategic Brief.

### What are the Suggested Key Support Tasks and Project Strategies?

Feedback reviews are an important task: they will help to ensure that the Project Outcomes reflect both the client's requirements and best practice. The process requires an understanding of a wide range of strategies, issues and approaches that will ultimately underpin the design information developed during the subsequent stages and the overall operation and performance of the project when 'In Use'.

Feedback reviews can include assessments of:

I Project Performance
I Maintenance and Operational Strategies
I Post-occupancy Evaluations
I building performance evaluations
I Research and Development strategies
I performance against energy and sustainability targets.

Understanding how these approaches performed on projects of a similar use, type and scale may influence the overall design approach and the strategic performance targets, outlining what 'best practice' considerations should be applied. Reviewing these strategies can also help to define the key Project Outcomes in terms of quality, performance and environmental aspirations.

The iterative nature of the information review process, of learning from the end to define the beginning, is critical to the structure of the RIBA Plan of Work 2013, allowing successful objectives, tasks and information to be measured and assessed.

### Project Outcomes

Using Project Outcomes to develop the brief represents a shift from physical, functional and budgetary requirements, to a focus on how a project can enhance aspirations, experiences, operation and performance for the client and users. Successful outcomes might include improvements to outputs and productivity in commercial environments, increased turnover and sales for retail schemes or improved personal development within educational facilities.

*The Value Handbook: Getting the most from your buildings and spaces*, produced by the Commission for Architecture and the Built Environment (CABE), outlines six specific value definitions:

– Exchange
– Use
– Image
– Social
– Environmental
– Cultural

These are detailed in table 0.1.

## Project Outcomes (*continued*)

*Table 0.1  Types of value*

| TYPE OF VALUE | WHAT DOES IT MEAN? | HOW IS IT MEASURED? |
|---|---|---|
| **Exchange value** | The building as a commodity to be traded, whose commercial value is measured by the price that the market is willing to pay. For the owner, this is the book value, for the developer the return on capital and profitability. Also covers issues such as ease of letting and disposability. | Book value<br>Return on capital<br>Rental<br>Yield |
| **Use value** | Contribution of a building to organisational outcomes: productivity, profitability, competitiveness and repeat business, and arises from a working environment that is safe in use, that promotes staff health, well-being and job satisfaction, that encourages flexible working, teamwork and communication, and enhances recruitment and retention while reducing absenteeism. | Measures associated with occupancy, such as satisfaction, motivation, teamwork. Measures of productivity and profitability, such as healthcare recovery rates, retail footfall, educational exam results, occupant satisfaction. |
| **Image value** | Contribution of the development to corporate identity, prestige, vision and reputation, demonstrating commitment to design excellence or to innovation, to openness, or as part of a brand image. | Public relations opportunities<br>Brand awareness and prestige<br>The recognition and 'wow' factors. |
| **Social value** | Developments that make connections between people, creating or enhancing opportunities for positive social interaction, reinforcing social identity and civic pride, encouraging social inclusion and contributing to improved social health, prosperity, morale, goodwill, neighbourly behaviour, safety and security, while reducing vandalism and crime. | Place making<br>Sense of community, civic pride and neighbourly behaviour<br>Reduced crime and vandalism. |
| **Environmental value** | The added value arising from a concern for intergenerational equity, the protection of biodiversity and the precautionary principle in relation to consumption of finite resources and climate change. The principles include adaptability and/or flexibility, robustness and low maintenance, and the application of a whole life cost approach. The immediate benefits are to local health and pollution. | Environmental impact<br>Whole-life value<br>Ecological footprint. |
| **Cultural value** | Culture makes us what we are. This is a measure of a development's contribution to the rich tapestry of a town or city, how it relates to its location and context, and also to broader patterns of historical development and a sense of place. Cultural value may include consideration of highly intangible issues like symbolism, inspiration and aesthetics. | Critical opinions and reviews<br>Professional press coverage<br>Lay press coverage. |

## Project Outcomes (*continued*)

Using these value definitions in association with operational performance requirements can inform the Strategic Brief, in addition to helping evaluate the overall success of the completed project.

Analysing and assessing all the different forms of Feedback will help to outline best practice guidance to particular building sectors and types; however, these forms of feedback should always be assessed and measured within the context of their specific project in order to ensure they do not adversely impact other strategic requirements and aspirations. Economies of scale, environment, context, brief, budget and performance should all be assessed to test the viability of best practice approaches; for example, solar control systems found to be exceeding performance targets on a small project might not be suitable or as efficient on a larger project within the same sector.

## Best practice information

Best practice information documents a solution or an approach that has consistently achieved or exceeded its desired outcome/ performance and can be used to develop a similar approach or as a benchmark against which to test and assess the performance of other systems.

## How can successful Project Outcomes be targeted?

Post-occupancy Evaluation (POE) and building performance evaluation (BPE) utilise direct operational and user experiences as a basis for assessing whether a project works as intended, achieving its desired outcomes. Understanding performance data, such as productivity, profitability and energy use, in relation to how the building is used can provide valuable information for improving existing systems and for developing the design, procurement and operation of future projects.

While key performance issues can be identified and resolved through POE and BPE, this process does tend to identify the areas of a project that have failed or are in the process of failing. Research has, for example, highlighted that thermal discomfort can tend to be associated with lower perceived workplace productivity and be a symptom of energy inefficiencies, badly configured control systems and poor management. Despite POE identifying problems, and even suggesting solutions, these negative insights can become a significant barrier to ensuring that appropriate reviews are implemented and, moreover, to the findings being made available to the team and other teams to learn from and develop.

Fears of any liability or litigation will need to be overcome when identifying any areas that will ultimately increase the performance of a facility and, more significantly, the knowledge and competency of the profession to effect future solutions.

### Methods of establishing POE and BPE

There are a number of different methods of evaluating system performance and user satisfaction. Examples of these include:

- BRE Design Quality Method (DQM): www.bre.co.uk
- Design Quality Indicators: www.dqi.org.uk
- Soft Landings: www.usablebuildings.co.uk
- Guide to Building Performance Evaluation:
  www.instituteforsustainability.co.uk/guidetobpe

The Soft Landings approach represents a significant review of the way the project team collects, analyses and uses operational and technical performance data from new-build and refurbishment projects. As a process it avoids many of the fears associated with highlighting poor areas of performance. For more information see the following publications from www.bsria.co.uk:

- BSRIA BG4/2009 *The Soft Landings Framework: For better briefing, design, handover and building performance in-use*
- BSRIA BG38/2012 *Soft Landings Core Principles*
- BSRIA BG 45/2014 *How to Procure Soft Landings: Guidance*

Access to use and performance data is invaluable for setting and achieving targets for future projects, and while some reliable information from existing studies is now available, Stages 0 and 7 recognise that a wider database and a cultural shift towards publishing these outputs will be required to ensure that information developed during this early stage benefits fully from the potential of Feedback.

### How can the delivery process contribute to the Strategic Brief?

The project delivery is typically reviewed and defined within later stages of the Plan of Work. However, there may be specific processes and procedures, on the client's side or otherwise, that have benefitted a previous project which can be used to inform and improve strategic outcomes at this stage. These lessons learnt can improve:

I procurement, manufacturing and construction strategies
I stakeholder needs and requirements
I key client procedures
I key project procedures
I process decision points
I key decision making.

Understanding procedures and process performance will also inform decisions relating to the way information is produced and managed, either through a Building Information Modelling (BIM) approach or through 2D and 3D CAD processes.

### How does CAD differ from BIM?

The production of drawn information has progressed with the evolution of computer systems, from drawing boards and ink to a predominantly computer-aided design (CAD) approach, which, put simply, has digitalised the drawing process, allowing faster and more accurate provision of information. Drawing in pen or using CAD is still intrinsically the same process, using a vector-based approach to describe the design. Building Information Modelling, although more of a process, describes the design through the placement of objects, replicating the actual solution in a modelled format. These objects can also contain information relating to the cost, programme, specification, finish and maintenance requirements, allowing all the information relating to a project to be developed, accessed and maintained within a single source.

## What information can constructed projects contribute?

Benchmarking employs actual metrics and measurable project-related research and data developed through precedent studies and the analysis of the outcomes from projects of a similar type and scale. The process measures and compares quantifiable variables, such as area, size of space, functional layout and cost/m$^2$, against a number of different parameters, including:

I industry best practice
I functional requirements
I performance requirements
I operational requirements
I energy use
I process implementation
I financial expenditure.

### Access to benchmarking data

A number of professional bodies provide online access to benchmarking data, either openly or through a subscription service, providing quick and easy access to up-to-date best practice information. Other industry-related sites also provide measured data in relation to the parameters outlined above. These include:

- www.carbonbuzz.org
- www.carbontrust.com/resources/faqs/sector-specific-advice/energy-benchmarking
- www.rics.org/uk/knowledge/bcis
- www.architecture.com/RIBA/Visitus/Library
- www.building.co.uk/data/cost-data/cost-model
- www.building.co.uk/buildings/technical-case-studies
- www.ajbuildingslibrary.co.uk
- www.cibse.org/building-services/building-services-case-studies
- www.breeam.org/case-studies.jsp
- www.bco.org.uk/Research/Best-Practice-Guides.aspx

Experienced clients may have completed projects from which they can offer benchmarked comparisons, and first-time clients might benefit from an understanding of how others have successfully achieved outcomes and aspirations, prior to determining their own brief.

This type of analysis typically uses completed construction information to visually express successful areas for clients to review. This allows both a qualitative and comparative understanding of specific precedents and is a good opportunity for using the information to define aspirations, outcomes and requirements within the Strategic Brief.

An extensive amount of data relating to specific parameters and functional requirements of building types is available on the internet. However, this information is often without context, incomplete and not validated. As highlighted above, some professional institutions and sector specialists record and validate data so that new projects can benefit from the benchmarked information, although both these forms of research should be substantiated using a number of alternative sources before being used to inform the Strategic Brief. In some instances, if locality is not an issue, it may be better to visit exemplary projects first hand and discuss with the owner/users how the building performs so that these insights can be reviewed and, if pertinent, captured within the brief.

### How is Feedback evolving?

There are other embryonic forms of feedback associated with information modelling systems, such as those emerging from some of the GSL Early Adopter projects outlined on the BIM Task Group's website (www.bimtaskgroup.org), where the opportunity for utilising a whole or a part of the digital model to form the brief for a similar project is being explored. These 'digital briefs' can comprise components that performed particularly well, or whole areas that demonstrate best practice in some spatial, functional or qualitative manner, that would be beneficial in the development of future designs. The HMYOI Cookham Wood Report (www.bimtaskgroup.org/reports) highlights the reuse of a prison cell, but this could also be a toilet pod, classroom, operating theatre or auditorium.

Additionally, because of the way performance data and information are integrated within BIM objects, these digital briefs can not only highlight the spatial metrics, but can also include the environmental and construction performance, the specification and the final costs. This creates an element of certainty within the strategic requirements that may inform the budget and whole-life costing aspirations.

Using modelled representations of successful performance and spaces as a brief will require a rigorous validation process in relation to context, use and the appropriateness and accuracy of any embedded information. With numerous BIM components, areas and spaces accessible online, care should be taken when assembling digital briefs to utilise only those from reliable sources, where the information has been independently checked against compliance and statutory issues, as a minimum.

While the standardisation of a cell or a cell block can be understood in the context of outlining specific sector requirements within a digital brief, there is a danger that some briefs might begin to digitally assemble whole buildings as a representation of required outcomes, without the appropriate consideration of adjacencies, site and context, or the rigour of a process and a functional overview. Ultimately, a common sense approach will be required when compiling this information to ensure it is relevant and appropriate.

Digital briefing will develop within sectors and building types to which it is particularly suited. If this approach is appropriately validated and utilised in context, the Strategic Brief might significantly change in the future, moving from a descriptive text-based document to an assembly of digital representations, performances and requirements defining the quality, outcomes and aspirations of the project.

## Documenting the Strategic Brief

The Strategic Brief will form the basis of both the Initial Project Brief within Stage 1 and the Final Project Brief in Stage 2. In order to facilitate these future stages appropriately the Strategic Brief will, subject to the project requirements, endeavour to cover the principles outlined in the checklist below in order to define the client's Core Objectives and outcomes.

### The Strategic Brief – information checklist

Typically, the Information developed within the Strategic Brief will include:

– the client's strategic objectives
– the client's constraints, requirements and specific needs

## The Strategic Brief – information checklist (*continued*)

- requirements for futureproofing
- other stakeholders and their needs
- a review any physical and operational constraints relating to probable sites
- a definition of the Project Outcomes

The information developed will document a broad understanding of:

### General

- the quality requirements
- how the success of the project will be measured
- the expertise required to deliver the project
- the method of information production and intended uses (BIM or CAD)
- any key Project Programme periods
- any key client project procedures

### Functional

- the overall areas and other spatial requirements
- any key departmental structures
- any specific technical requirements, such as servicing or operational constraints

### Environmental

- the strategic Sustainability Aspirations
- an overview of any context and constraints
- durability, lifespan and maintenance requirements
- flexibility and future uses

### Statutory

- any statutory requirements and constraints, including specific planning issues and pre-application advice requirements

### Financial

- cost appraisal studies and parameters
- the overall Project Budget – capital and operational expenditure

The document may also:

- review likely procurement strategies
- identify what is expected in response to the brief
- state what decisions are needed and from whom
- include sign-off procedures.

The project responses to these issues will be determined by the size and scale of the scheme and the client's requirements. It is important to consider carefully what areas are not needed at this stage to ensure that the Strategic Brief remains a simple, clear and concise document, delivering all the strategic information to progress the design.

## What are the Information Exchanges at the end of Stage 0?

The Information Exchange required at the Strategic Definition stage comprises:

I   the Strategic Brief.

This document should be reviewed, approved and signed off at the end of the stage.

### What are UK Government Information Exchanges?

In 2011, the UK government highlighted that it could significantly benefit from improvements in cost, value and carbon performance through the use of open, sharable asset information. It has subsequently produced a number of strategies to assist the implementation of BIM throughout the construction industry. These include:

I   the Digital Plan of Work (dPOW)
I   Employer's Information Requirements (EIRs)
I   plain language questions (PLQs)
I   Construction Operations Building Information Exchange (COBie)
I   a unified classification system
I   information protocols and standards.

The UK government has also mandated that all publicly procured projects and assets will achieve the requirements of a Level 2 Building Information Model, as defined by the Bew-Richards BIM maturity diagram (figure 0.1). This will provide a 'fully collaborative 3D BIM (with all project and asset information, documentation and data being electronic)' (Government Construction Strategy, 2011).

The mandate sets out a timeline to achieve this standard, as a minimum, by mid-2016.

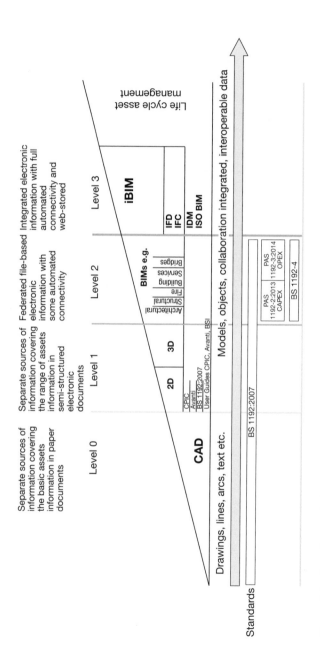

Figure 0.1  *Bew-Richards BIM level of maturity diagram*

The BIM Task Group website (www.bimtaskgroup.org) provides a detailed overview of the strategies, systems and initiatives developed in support of this mandate, most of which are now in place to support a collaborative working environment.

I Level 0: a mix of unmanaged CAD and paper-based information exchange systems.
I Level 1: managed CAD in a 2D or a 3D format with a collaborative tool, a common data environment (CDE), a standard method and procedure (SMP) and a standard data structure and format based on BS 1192:2007.*
I Level 2: a managed 3D environment created through collaborative working in separate discipline BIM tools with attached data using a developed CDE and SMP based on PAS 1192-2:2013.*
I Level 3: a fully integrated, collaborative process enabled using 'web services' and compliant with emerging Industry Foundation Class (IFC) standards.
I A Level 3 project will also accommodate web-based construction sequencing (4DBIM), cost information (5DBIM) and project life cycle management information (6DBIM).

## What is BIM?

BIM is an acronym that can be taken to represent either:

– building information model
– building information modelling, or
– building information management.

The use of these definitions remains fairly interchangeable and although consensus is beginning to settle on the model definition, its application is more accurately reflected when described as a wider process, such as building information management.

The RIBA Plan of Work 2013 uses the acronym to mean 'building information modelling' (see page 229).

However it is defined, BIM is more than just a modelling tool. It should be considered as a *process* that facilitates the delivery of a common output through collaboration and coordination.

*Refer to Stage 1: Preparation and Brief for a detailed overview of these standards.

## What is BIM? (*continued*)

### Why is BIM different?

A BIM model is typically produced by a single discipline, eg the architectural model (AM) or the structural model (SM), and integrated with other design models to create a single representation of the project. This federated 3D model allows the full design to be accessed and viewed by the whole project team, including the client, users and stakeholders.

PAS 1192-2:2013 refers to these models, federated or otherwise, as the Project Information Model (PIM), comprising graphical and non-graphical data and documents compiled throughout the design and construction stages. Once this information is verified as 'as built' the PIM is finally developed into the Asset Information Model (AIM) for assisting the operation and maintenance of the project during the In Use stage.

A PIM is not drawn as with CAD systems, but rather assembled from a series of modelled components. These generic or very specific objects enable geometry and information to be coordinated and reviewed against other similar building elements and systems within the design. The objects can be developed to inform:

- building geometry
- spatial relationships
- geographical information
- component quantities
- component properties
- specifications
- cost information
- buildability
- programme
- schedules
- maintenance requirements.

BIM represents a significant shift from traditional CAD methods, not only in the composition of the design, but also in both working methods and the transfer of information, with the whole project team sharing and utilising information within the context of the project model. This workflow can reduce the risk of coordination errors, and when additional clash detection software is used, issues within the developing design can easily be identified and resolved.

Using a federated model to provide all Project Information digitally defines a Level 2 approach, as highlighted within the Bew-Richards level of maturity diagram.

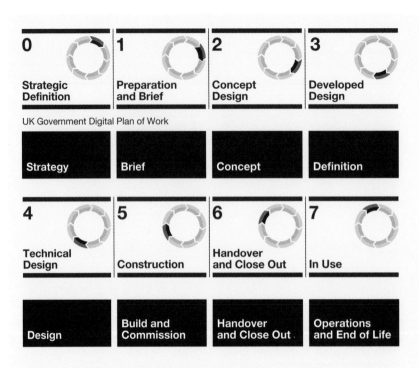

| 0 | 1 | 2 | 3 |
|---|---|---|---|
| Strategic Definition | Preparation and Brief | Concept Design | Developed Design |

UK Government Digital Plan of Work

| Strategy | Brief | Concept | Definition |
|---|---|---|---|

| 4 | 5 | 6 | 7 |
|---|---|---|---|
| Technical Design | Construction | Handover and Close Out | In Use |

| Design | Build and Commission | Handover and Close Out | Operations and End of Life |
|---|---|---|---|

Figure 0.2    RIBA Plan of Work 2013 and Digital Plan of Work comparison

The plain language questions (PLQs) outlined by the BIM Task Group were created, using specific project examples, to set out the client's minimum information requirements necessary to fulfil the Employer's Information Requirements (EIRs) for each particular stage. The Digital Plan of Work (dPOW) highlights a number of Data Drops (specific information exchanges), which are defined by the model requirements and the resolution of each stage's particular PLQs.

All dPOW stages correspond to the RIBA Plan of Work 2013 stages, as highlighted in figure 0.2.

There are no UK Government Information Exchanges (Data Drops) required at this stage. However, there are a number of PLQs, which require responses to define:

I   the information management strategy

- standards and protocols
- an overview of the possible development options
- the characteristics of the site
- an understanding of key technical requirements
- any performance benchmarks and targets
- an approach to standardisation
- stakeholders' requirements
- capital and operational costs
- precedents and lessons learnt.

Although these questions are specific to government-procured projects, they can be used as a starting point to help outline the minimum requirements for information produced within this and subsequent stages in order to successfully progress the design.

### Plain language questions

The full list of standard PLQs used to develop EIRs in conjunction with the dPOW can be reviewed at www.thenbs.com/BIMTaskGroupLabs/questions.html

### Standards and protocols

The protocols and standards developed by the UK government underpin a 'push' and 'pull' strategy for adopting and developing Level 2 BIM processes within the construction industry.

The dPOW and the unified classification system are the only two processes that have yet to be finalised in support of the overall BIM strategy. These are currently being developed within a BIM Toolkit produced by NBS, in conjunction with a number of other BIM-focused organisations.

The BIM Toolkit will be freely accessible during 2015 and will help to 'define, manage and validate responsibility for information development and delivery at each stage of the asset lifecycle'. In addition to supporting all the current key standards and guidelines, the outputs from the BIM Toolkit will also assist the development and production of the brief, Schedules of Services and the Design Responsibility Matrix.

### Who produces the Strategic Brief?

The Strategic Brief represents the beginning of a project's development. While it is not uncommon for experienced clients to use their own in-house professional team to outline their Business Case and initial operational requirements, it is likely that most clients will be less sophisticated and will require the input of an architect or other design team member to help develop and articulate the required information. This assistance could also be provided by an independent consultant, such as an RIBA Client Advisor, who will provide impartial advice on critical issues and assist in developing the strategic appraisal and brief for the project.

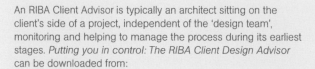

### RIBA Client Advisor

An RIBA Client Advisor is typically an architect sitting on the client's side of a project, independent of the 'design team', monitoring and helping to manage the process during its earliest stages. *Putting you in control: The RIBA Client Design Advisor* can be downloaded from:

www.architecture.com/Files/RIBAProfessionalServices/Directories/ClientDesignAdvisor.pdf

Depending on the nature of the project, the client in conjunction with the project manager and the lead designer/design adviser might also set out how the project team will be structured, what the scope of work for each team member might be and when they will be appointed.

### The project team

*Assembling a Collaborative Project Team* (by Dale Sinclair, RIBA Publishing, 2013) suggests that those elements outlined in figure 0.3 will in many cases comprise the basis of the project team. In addition it highlights that cultural changes in the way the construction industry procures and delivers projects, better informed clients, the early involvement of the contractor and the expanding scope of the design team have resulted in a shifting dynamic that will need to be reviewed on a project-by-project basis in order to develop a coherent approach with the project team.

**The project team (*continued*)**

*Figure 0.3   The structure of the project team*

Some clients, in order to minimise design costs, may not want to appoint numerous consultants for the early design stages, prior to a planning permission being secured. The risks associated with this approach should be clearly outlined within the brief, so that any significant changes occurring as a result of delayed specialist or professional input can be addressed accordingly.

Should the project implement a BIM approach at this stage, it is essential to involve the core design consultants early on, to assess and test the key constraints and outcomes, so that the Project Information is progressed using a collaborative approach.

## Chapter summary 0

Stage 0 is a new stage, during which project-specific needs are defined based on organisational requirements and on knowledge derived from the client's ambitions and from successful projects of a similar type and scale.

While the Strategic Brief is principally derived from these client-specific requirements, such as the Business Case and stakeholder input, Feedback is fundamental to understanding performance and outcomes from similar projects and how achievable aspirations and targets can be defined and delivered from the outset. Being able to identify theses core objectives succinctly at this stage enables a thorough appraisal to be developed and reviewed, ultimately determining whether the project progresses.

The data and information produced and gathered from projects in operation is key to structuring the overall strategic response. Any improvements that can be identified by looking at how buildings have been produced and operate may result in significant changes to the way subsequent strategic aspirations are defined, with both the constructed and operational information becoming essential to the overall process.

The Stage 0 Information Exchange is:

| Strategic Brief.

Additional information reviewed might include:

| Business Case
| Site Information
| appointment documentation
| Feedback, research and benchmarking data
| potential design team requirements.

# Preparation and Brief

# Chapter overview

Stage 1: Preparation and Brief is a decisive stage within the overall project process, during which the Strategic Brief is tested against early ideas and responses using Feasibility Studies, Site Information and research, and the suggested scope, approach, responsibilities and services required to deliver the appropriate information are outlined within the Initial Project Brief.

**The key coverage in this chapter is as follows:**

Producing the Initial Project Brief

Developing Feasibility Studies

What are the Information Exchanges at the end of Stage 1?

Strategies supporting the Initial Project Brief

Structuring Project Information

Understanding roles and responsibilities

# Introduction

While the Initial Project Brief represents the primary information exchanged at the end of Stage 1, there are a number of key tasks and activities completed throughout the stage that will fundamentally inform the design process. Ideally, these should also be finalised prior to progressing to Stage 2.

These two key work streams can be loosely defined as information gathering and information planning. Information gathering typically covers:

Site Information, including surveys

Feasibility Studies, research and analysis

precedent studies

brief development

Cost Information.

Information planning will develop:

the Project Execution Plan

the Technology and Communication Strategies

the Design Responsibility Matrix

the Schedules of Services.

Understanding what information will inform and assist the development of the Concept Design is central to the development of the Initial Project Brief, as is the way the data will be formatted and exchanged. The standards and protocols used to produce the project data will facilitate the accurate transfer of information between team members. It is therefore important that these are clearly understood as they may affect the development and communication of ideas later within the project.

The outcomes and aspirations identified within the Strategic Brief will have suggested the likely key project team members required to develop the design. This stage allows for the full Schedules of Services to be finalised, based on the emerging brief and feasibility studies. The Design Responsibility Matrix outlined in the RIBA Plan of Work 2013 Toolbox (www.ribaplanofwork.com) and the Digital Plan of Work BIM Toolkit (https://toolkit.thenbs.com) can be utilised to highlight the specific Project Information requirements and who will be responsible for delivering them during each stage. If particular information or disciplines are not appointed or delivered, either in full or in part, until a later stage, the lead designer should ensure that all the implications and risks are reviewed and understood before finalising the project scope.

## What are the Core Objectives of this stage?

The Core Objectives of the RIBA Plan of Work 2013 at Stage 1 are:

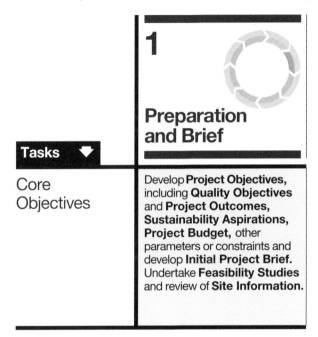

The Core Objectives support the development of the Initial Project Brief and are progressed from a review of the Strategic Brief, the development of the Project Objectives, including the Project Outcomes, and Feasibility Studies to ensure the brief and site are aligned. They are set out as outlined in the following sections.

## Producing the Initial Project Brief

The Initial Project Brief will provide a coherent record of the Site Information and Feasibility Studies and the key project decisions in a format that can be reviewed by all those who will commence design work at Stage 2. It will also clearly identify the Project Outcomes and Project Objectives and the direction the design should follow within subsequent stages.

The Initial Project Brief should consider:

l the validation of the overall scope and purpose of the project
l site or building survey and studies
l research into functional needs and accessibility audits
l environmental impact considerations
l statutory constraints
l Feasibility Studies
l cost appraisal studies
l functional requirements
l Project Programme and phasing review.

**How does the Initial Project Brief differ from the Strategic Brief?**

The Strategic Brief outlines the overall vision within the Project Objectives and Project Outcomes in relation to the Business Case. The Initial Project Brief refines this 'high level' information, identifying what the key project decisions are, when they need to be taken, who will be required to develop the design and what the project team's overall scope might be.

### The Initial Project Brief – information checklist

Typically, the information developed within the Initial Project Brief will include:

General
– the client's Project Objectives and Project Outcomes
– quality standards, and a suitable measuring system to assess performance against outcomes
– the Sustainability Aspirations
– a review of the context, including site history, topography and geology
– the defined site boundary
– studies commissioned to date

## The Initial Project Brief – information checklist (*continued*)

- plans for future expansion
- likely stakeholders, including statutory bodies, eg English Heritage
- initial project risks

### Procedural
- Contractual Tree and Project Roles Table
- project roles and the project team required to carry out the development
- the Project Execution Plan summary

### Functional
- detailed functional requirements of direct client/user client
- preferred spatial relationships and orientation
- adjacencies and spaces that require separation
- end use considerations
- internal and external environmental requirements and conditions
- specific access requirements, including inclusive access
- life expectancy of building and components
- any additional project/use-specific requirements

### Environmental
- below-ground services and development constraints
- likelihood of archaeological or antiquarian discoveries
- future infrastructure or other development plans
- geological conditions, hazardous substances, presence of contaminated land
- existing buildings, impacts and conditions
- outstanding Site Information requirements

### Statutory
- specific planning constraints
- constraints arising from previous consents or conditions
- the likelihood of environmental or traffic impact assessments
- planning issues within the locality
- the impact of the local planning requirements
- conservation areas and listed buildings
- stakeholder interests, party walls, rights of light, access and easements

### Financial
- the Project Budget
- funding or institutional requirements or restrictions
- benchmarking and approximate £/m$^2$
- operating expenditure, costs in use and other whole-life costs.

The information documented should be appropriate to the project scale and size. Too much information can obscure the overall intent, so it is important to consider what might be omitted from the brief for clarity. Some projects may require detailed reports under the headings outlined above, in which case an executive summary can be used for setting out the more significant issues, ensuring the most important information is instantly accessible. Appendices and supporting documents can also be used for larger reports.

The final format of the Initial Project Brief and any other project documentation should be considered from the outset. Structuring the key project documents in a consistent way will avoid reproducing and reworking information for use elsewhere and will allow it to be presented and represented in a coordinated and coherent format.

### Formatting information

Careful consideration should be made when deciding on the appropriate software to collate this information. Different design disciplines often use different document publishing and spreadsheet software, which can make the overall compilation very time consuming and unproductive. The lead designer will be responsible for collating the coordinated information at Stage 3 and their preference for the format and presentation software and the design team's ability to produce information accordingly should be determined at this stage to assist in the transfer and communication of key information here and in the later stages.

## Developing Feasibility Studies

Feasibility Studies help to define the requirements of the Initial Project Brief, which must outline the project approach in relation to specific sites, the composition and requirements of the project team and, ultimately, what the project team will produce and when.

The vision, Project Outcomes and key functional requirements outlined within the Strategic Brief will provide the basis for testing the detail of the Initial Project Brief, reviewing:

I Site Information, context and constraints

   | key project fixes
   | use and adjacencies
   | spatial requirements
   | benchmarking and research.

These early studies may highlight that the preferred location is not the most suitable or that the client's aspirations can be achieved using a more efficient solution than that envisaged. Moreover, the site location and orientation might significantly influence any preferred design location in favour of delivering a more efficient and sustainable approach. The key criteria and decisions developed from these studies should be documented, outlining the viability of each option as well as any areas of information and work required to progress and test each solution. The potential costs can also be reviewed and compared against the Project Budget to determine the relative value of each approach.

### Feasibility Studies – information checklist

The Feasibility Studies might include:

- Summary of the Strategic Brief
- Issues for the Initial Project Brief
- Design statement, setting out design objectives in relation to:
  - external environment and landscape
  - functional requirements
  - key constraints
  - delivery, flexibility and adaptability
  - aesthetics, identity and brand
  - structure, environmental performance and sustainability
  - access, safety and security
  - materials, durability, maintenance and value
- Summary of stakeholder consultations
- Town planning report
- Survey requirements
- Environmental report
- Budget and cost analysis
- Summary of Project Execution Plan: key Project Strategies
- Risk register.

This stage represents the starting point for the creation of the Project Information. The project team should ensure that all the data can be reviewed, validated and reused efficiently by other team members when exchanged, especially given that the procurement strategy chosen will affect the Contractual Tree for the preparation of the Concept Design.

## What Site Information is required?

At the outset of this stage the project team may have already accumulated a significant amount of Site Information from the client and other sources. The full extent and format of the information available needs to be collated, reviewed and validated in order to identify whether it is of any use, if there are any omissions and where further research and information may be required. Understanding the specific site and project requirements will help to identify any statutory, geotechnical, existing building and topographical constraints as well as what further Site Information might be required to inform the design.

There are a number of areas that should be explored to identify the project Site Information requirements and provide a comprehensive overview of the existing conditions. These include:

| existing building conditions, uses and drawings
| existing services and utilities
| below-ground obstructions
| local amenities use and location
| boundaries, land registry documents, drawings and site ownership
| site and planning history, including:
  ○ existing consents and conditions
  ○ environmental impact assessment requirements
  ○ traffic impact assessment requirements
  ○ existing covenants
  ○ easements
  ○ rights of way
  ○ party wall issues
  ○ rights to light
  ○ heritage issues
  ○ historical context and listings
  ○ key views
| existing constraints, hazards and contamination on or around the site
| access to the site, pedestrian, vehicular and public transport network
| ecological and geological conditions
| trees and hedges (including Tree Preservation Orders)

| local climate
| seasonal variations and flood risk.

Reviewing these areas provides an opportunity for the project team members to familiarise themselves with the site and its context and will help to identify any requirements for additional detailed surveys.

It is important to identify what survey information is needed as early as possible, how it should be produced and who would be best placed to provide it. Some clients will be able to procure surveys from consultants they will have engaged previously; alternatively, the project team will need to define the scope and survey requirements clearly for the client to procure tenders and proposals from suitable surveyors. Following the development of the scope it may be beneficial to walk the site with the preferred surveying consultants to understand the key requirements and areas where additional information may be required. Any resulting omissions and additions should be added to the final scope, which should also identify the level of detail required and the format of the survey information, the appropriate standards and protocols. The information should be geo-located, in accordance with the emerging Technology Strategy.

### Scoping surveys

The scope for site surveys can cover a number of areas, including:

**Existing buildings**

- measured
- structural
- condition
- demolition

**Topographical**

- land levels and formations
- below-ground obstructions
- utilities: gas, electricity, telecommunications, water, surface and foul drainage
- geological and geotechnical conditions
- unexploded ordinance

**Environmental**

- land contamination
- hazardous material
- ecological
- flood risk
- air quality
- acoustic

**Contextual**

- Archaeological
- Traffic and transport
- Photographic
- Previous uses
- Existing boundaries.

3D surveys can typically be viewed and used within standard CAD software, but they can also be provided in a 'flattened' format for use in a 2D environment. Subject to the project requirements, programme and cost constraints, it may be beneficial to procure a 3D survey and distil the information into a 2D file, if that is the required format for the project. The 3D view will be maintained and can be used to help the project team understand the layout of the site, even if the final information is utilised in a linear format.

## Formatting surveys

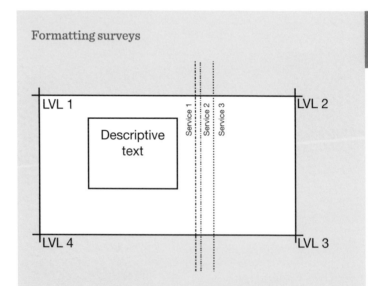

*Figure 1.1a    Layered survey information, plan view*

Figure 1.1a shows how layered survey information might appear when viewed on screen and figure 1.1b shows it in a more conventional plan view. By separating the annotated information from the geographical data, the structure of the file allows the features of the site to be reviewed in an uncluttered format, while the actual data can be used and viewed in plan as a visual check. Also, if the modelled information is geographically located in relation to the agreed project datum, each point within the file will correspond to its associated annotated value, allowing levels to be understood without necessarily needing the 2D notes and values.

Regardless of which final format is required for the information, ensuring that the data is structured appropriately and in accordance with the project standards is essential to the progression of future design information. When scoping a 3D survey, it is beneficial to specify that text and numerical data are separated vertically from the actual model representation.

Ensuring that the survey information is set out in accordance with the project requirements will avoid unnecessary reworking by the project team and will significantly reduce the risk of information becoming corrupted or inadvertently amended or deleted when being adjusted for clarity of view.

### Formatting surveys (*continued*)

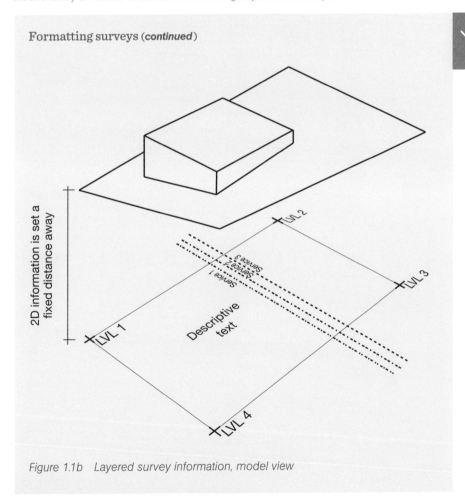

Figure 1.1b   Layered survey information, model view

It is also possible to specify the use of point cloud technology to provide surveys of existing site conditions and buildings. This process is relatively easy to complete and its cost is becoming more comparable with that of other 2D and 3D survey methods. Point cloud data can be converted into a BIM-compatible file format for use within a Project Information Model (PIM) and can also be supplemented using colour 3D photography to help identify specific services or elements within the recorded data.

### Point cloud information

Laser technology can be used to automatically take a large number of measurements from the surface of an object. The process creates a digital cloud of geometrical points, the point cloud, which can be viewed as a 3D representation of what has been scanned. A 3D CAD/BIM model of the existing buildings or site conditions can then be created from the point cloud data for use in editing software. It is important to always specify whether the conversion process is required.

For more information on using point clouds refer to *BIM in Small Practices* by Robert Klaschka (RIBA Publishing, 2014), Case Study 07: Using existing building BIM, and www.thenbs.com/topics/bim/articles/pointCloudSurveys.asp

## What are the Information Exchanges at the end of Stage 1?

As with Stage 0, the Information Exchange for Stage 1 focuses on a single output:

I   the Initial Project Brief.

This stage also has an equivalent exchange within the UK government's Digital Plan of Work (dPOW):

I   UK Government Information Exchange: Data Drop 1.

Data Drop 1 represents the first required exchange of developed information in the dPOW process and comprises a modelled response to the plain language questions (PLQs) addressed throughout this and the previous stage. The information should be sufficient to deliver an initial Construction Operations Building Information Exchange (COBie) output to validate its development, to ensure that the design and Initial Project Brief are aligned and to allow the outline Business Case to be approved.

### What is COBie?

The COBie output identifies and collates the non-geometric information that is needed to exchange managed asset information over the life of a project (figure 1.2).

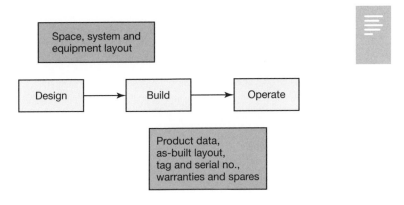

Figure 1.2    COBie basic workflow

PAS 1192-3:2014 defines an asset as an 'item, thing or entity that has potential or actual value to an organization', noting also that:

An asset may be fixed, mobile or movable. It may be an individual item of plant, a system of connected equipment, a space within a structure, a piece of land, or an entire piece of infrastructure or an entire building or portfolio of assets.

The information is extracted from the BIM model in a simple spreadsheet format that can accommodate large and small project models, allowing the team to document specific information both spatially, by floors, sectors and zones, and physically, by actual product types, components and systems.

While the process is set out to manage the constructed elements of the project during the In Use stage, the dPOW requires COBie data to be documented as it is created within each stage of the design, construction and commissioning process. This approach simplifies the work required to record and document the data at the completion of the project.

### Standards and protocols

BS 1192-4:2014 *Collaborative production of information. Fulfilling employer's information exchange requirements using COBie. Code of practice*
This specification focuses on the client's information expectations and requirements for each stage of the project and how COBie can assist in its production.

A crucial point is that the dPOW assumes that projects will follow a procurement route that allows the earliest involvement of the contractor and its supply chain. To give them the best possible information at the start of Stage 2, Data Drop 1 includes a BIM model derived from the Feasibility Studies containing spatial allocations, environmental parameters and any other constraints.

During Stage 1, the PLQs in support of the modelled Data Drop can be used to demonstrate an understanding of:

I design team scope
I brief development and sign-off
I BIM strategy
I Site Information, including physical constraints and utilities
I Project Budget outline and breakdown
I life cycle costs
I approach to planning consultations
I feasibility studies and the assessment of adjacencies and functional requirements
I facilities management and In Use requirements and strategies:
  o servicing strategies
  o information standards and protocols
  o construction sequencing and logistics.

## Strategies supporting the Initial Project Brief

While the development of the brief will use the defined Project Outcomes and Project Objectives to inform the design concepts, the Suggested Key Support Tasks and Project Strategies will support the way in which the information is structured, managed and progressed.

The Stage 1: Preparation and Brief Key Support Tasks include the preparation or agreement of:

| Project Execution Plan
| Handover Strategy
| Risk Assessments
| Technology and Communication Strategies
| Design Responsibility Matrix
| Schedules of Services.

### What information is required within the Project Execution Plan?

The need for project protocols and processes will depend on the size and complexity of the project and the number of disciplines involved. However, even on a small project a Project Execution Plan will be a valuable document. The plan should develop:

| outline contact information
| documentation of contractual relationships (Contractual Tree)
| communication protocols (see page 55)
| activities and how they are to be carried out (Project Programme and Technology Strategy, see page 55)
| sign-off procedures
| project progress, with any amendments and/or revisions, as required, submitted for acceptance before implementation.

The size of the project may affect the detail within the Project Execution Plan, which should ideally be completed as the design progresses so that it will capture and communicate the necessary information.

The Project Execution Plan should be developed early on within the Preparation and Brief stage to record any key project decisions, and also to outline specific areas that may affect the project scope, fee

## Project Execution Plan – information checklist

A typical Project Execution Plan will contain the following:

**Project description** (summarised from the strategic and initial brief documents)
- the client's design requirements
- an outline of the key brief elements
- any statutory and site-specific constraints
- an overview of the Sustainability Strategy
- procedures for controlling and reviewing the brief and the Plan
- summary of any significant risks

**Project organisation** (summarised from the strategic and initial brief documents)
- the identity of the client and representatives
- the identity of any key stakeholders
- agreed procedures for consultations/approvals
- project team members and their defined responsibilities and contact details
- summary of procurement approach

**Project controls**
- the Design Responsibility Matrix
- the Communication Strategy summary (with full strategy appended)
- the Technology Strategy summary (with full strategy appended)
- Information Exchanges – outline of information required to communicate the ideas
- the Project Programme

**Project development** (reference only, typically added at Stage 2)
- additional design management procedures
- the Health and Safety Strategy
- the Maintenance and Operational Strategy
- the Construction Strategy

**Change control** (typically added at Stage 3)
- project definition of change
- agreed Change Control Procedures for amendments to approved designs

### Project Execution Plan – information checklist (*continued*)

In Use processes (developed with information from the Handover and
Maintenance and Operational Strategies)
- scope for Feedback
- review of any Post-occupancy Evaluation
- review of Project Performance
- success of Project Outcomes
- areas for further Research and Development
- summary of 'As-constructed' Information.

and management requirements. Managing the development of the
design will ultimately ensure that, irrespective of the composition of
the project team and however it might change over the course of the
project, the support systems and procedures are clear, concise and
easily operated.

### Project Execution Plan requirements

For smaller projects, simply producing one or two lines
documenting how the design will be developed, when information
is to be produced and where it will be located, will benefit the
progression of the works. So too will producing a record during
the Construction and In Use stages of how the project was
delivered and how it performed against the original outcomes and
requirements.

### Developing a Handover Strategy: what allowances should be made for the project 'In Use'?

In addition to preparing the Information Exchanges required for construction,
it is also important to form an understanding of how design information
will be used during occupation and operation. Such requirements are
typically outlined within the Handover Strategy and Employer's Information
Requirements (EIRs), which, in addition to highlighting issues relating
to project completion, phased handovers, commissioning and other

post-occupancy activities, will set out what information the client requires for the In Use stage, perhaps as part of a Soft Landings or similar process. Understanding how these strategies will be developed will enable the project team to make allowances within the initial scope, fee and appointment for post-occupancy services, the restructuring of any 'As-constructed' Information and the appropriate management of modelled information 'In use'.

> **How the Project Execution Plan can be used to manage project risks**

Issues that are likely to affect the progress of the project should be identified as early as possible within the process, recorded and maintained within stage Risk Assessments. Assessments made at this early stage should be considered as an opportunity to acknowledge project constraints, to recognise objectives and priorities and to identify and mitigate the impact of any critical areas within the emerging ideas.

These initial reviews will be based on known site constraints and information developed within the Feasibility Studies, but it is essential to continually review any likely risks on receipt of additional and developed Site Information to understand if any have been mitigated. The management of remaining risks can be allocated to an appropriate project team member to develop a suitable response.

Should the project develop without this specific advice – for example, the client may wish to keep the design team to a minimum, to avoid a sizeable fee during these initial stages – specialist advice may be delayed. These risks should also be identified within the Risk Assessment as they may result in considerable reworking of the design and additional costs within subsequent stages.

## Structuring Project Information

The way we produce and utilise information on projects is constantly changing, influencing the way we communicate within the project team. The standards, processes and protocols used to facilitate these collaborations need to be determined as early as possible to ensure the accuracy and efficacy of the information produced and its proper management throughout the construction and occupation of the design. The Project Execution Plan outlines a number of different strategies to control these processes so that, even on small projects, the team can ensure that the right information is validated and communicated at the right time.

The Communication Strategy will outline the 'who, what and how' procedures that relate to information being discussed, issued, reviewed and checked, while the Technology Strategy will define the way information is created and managed within the project team.

### How should information be managed?

The Communication Strategy should typically outline protocols for:

I meetings
I workshops
I correspondence, including emails
I verbal communication
I information requests
I drawing comment processes
I reports.

Many of the above will be outlined within each discipline's quality assurance (QA) system. The Communication Strategy for the project may therefore:

I adopt the best protocols from the different QA systems
I allow each discipline to maintain its own procedures, or
I override the individual disciplines' procedures, in order to achieve a common approach.

If a definitive QA system is not outlined then the RIBA Quality Management Toolkit can provide a good starting point for controlling the key project processes.

## The RIBA Quality Management Toolkit

The RIBA Quality Management Toolkit comprises:

- *RIBA Quality Management Toolkit Overview*
  Outlines the philosophy behind the Toolkit approach and operation.
- *RIBA Project Quality Plan for Small Projects* (PQPSP)
  Highlights the processes for running individual projects –
  incorporating a simple, single-sheet short form for small projects,
  consultancies and investigations.
- *RIBA Quality Management System – Guidance*
  Over two sections, the guidance outlines the use and operation of
  systems that comply with the requirements of the international
  standard BS EN ISO 9001:2000:
  - Topic Guidance, which explains the benefits of having a quality
    management system
  - Documentation Tasks, which covers the necessary documentation
    and approaches.
- *RIBA Quality Management System – Quality Manual*
  The Quality Manual is a declaration highlighting commitment to the
  quality management system.
- *RIBA Quality Management System – Procedures Manual*
  Provides essential procedures covering both office and project
  processes. It includes templates and more detailed work
  instructions.

## How are different information production systems integrated?

The Technology Strategy sets out the hardware, software and format
protocols to be used by the project team in developing and coordinating
the design.

Information production within the construction industry is currently going
through a transitional period, developing from the more traditional 2D/3D
CAD drawings and documents into a predominantly data-driven modelling
environment. The Technology Strategy, especially during this transitional
phase, is a critical document for defining the structure and format of
the information, how it will be managed and how it will be developed,
ensuring that any misunderstandings that could arise as a result of the
production method or configuration are avoided.

Understanding the software and processes utilised by each discipline, including the differences and their implications, will ensure that software does not become a barrier to collaborative working. Issues relating to the hardware and software used should be considered prior to any design work commencing. However, as the design progresses and information from specialist designers, subconsultants and contractors is integrated, a review of these strategies, ideally at the beginning of each stage, will be required.

The procurement strategy adopted may also mean that the project team will change at specific stages. If this is the case, the structure of the information should be developed to ensure seamless exchanges throughout the course of the project.

### Why use design protocols and CAD standards?

The increased uptake and use of CAD has led to design consultants producing numerous single-party documents and standards that locally define and control the production of information. Given the differences within these standards and the lack of a common approach, coordinating the different disciplines has become increasingly problematic.

BS 1192:2007 *Collaborative production of architectural, engineering and construction information* outlines a number of standard methods and procedures to structure and manage project information. These protocols allow information to be created accurately and collaboratively, minimising any errors and delays associated with poor input and management. This standard has also recently been developed and expanded to support BIM processes outlined within:

I   PAS 1192-2:2013 *Specification for information management for the capital/delivery phase of construction projects using building information modelling.*
I   PAS 1192-3:2014 *Specification for information management for the operational phase of assets using building information modelling.*
I   BS 1192-4:2014 *Collaborative production of information. Fulfilling employer's information exchange requirements using COBie. Code of practice.*

It is important at this early stage to ensure that the Technology Strategy outlines that any information produced on the project, whether CAD or

BIM, maintains the collaborative requirements and standards set out in these documents, including as a minimum:

I a coordination strategy
I standards
I file naming conventions
I layer naming conventions
I geospatial location of the project
I exchange protocols.

## Where should digital protocols and standards be outlined?

The Technology Strategy will document the standards and procedures that the project information will be developed from, and any specific CAD or BIM protocols will be outlined or appended.

The 2D/3D CAD standard or the BIM execution plan (BEP) should ideally be used to form the basis of the Technology Strategy. The adoption of a coordinated and consistent approach will maximise production efficiency, while the standards and best practices set out will ensure the delivery of high-quality and uniform information.

### BIM execution plan

A BIM execution plan (BEP) will include:

– responses to any EIRs
– the scope of any BIM, including the key model uses throughout the project planning, design, construction and operational phases
– the processing and flow of information
– the required information exchanges and BIM deliverables
– a summary of the Communication Strategy procedures, IT equipment and QA requirements to support the implementation of BIM
– the authorised uses of the PIM created on the project
– the discipline responsibility for the development of each model element to a defined level of definition (LOD) for each stage of the project.

The Technology Strategy should be considered by all members of the project team and viewed as a live document; it should continually be updated via a series of reviews and workshops. The document should be revisited at the beginning of each stage.

A number of CAD standards and BEPs have already been developed within the industry.

The Construction Industry Council (CIC) has produced a suite of documents in support of the government's BIM objectives, including a BIM protocol template that can help to define the contractual and legal requirements of a project BIM model when appended to the main contract. It can also be used by the project as a starting point for setting out the BIM requirements by adding project-specific details to the appendices:

| Appendix 1: the Model Production and Delivery Table (see figure 1.5) includes references to the information required by the employer at each project stage
| Appendix 2: Information Requirements details the information management standards that will be adopted on a project.

### CIC BIM Protocol

The CIC's *BIM Protocol* can be downloaded from http://cic.org.uk/publications

### How should LOD be defined?

Whether a project is developed using 2D/3D CAD or through a BIM approach, it is important to understand what level of definition (LOD) will be achieved by the Project Information at the end of each stage and what that data can be used for.

LOD is an acronym developed primarily in support of BIM data, although it can also be used within a CAD environment. While seeking the same objective, there are a number of recognised interpretations for LOD. In the US, where the term was developed by the American Institute of Architects, it stands for 'level of development', and uses numerical definitions (LOD100, 200, 300 etc.) to describe the graphical content of components within a design. These are explained further within the BIM Forum's recently published *Level of Development Specification*.

In the UK, further to the release of PAS 1192-2, and the development of the dPOW, LOD is more commonly used to describe the 'level of definition'.

LOD has also been used to represent the 'level of detail' or 'level of design' contained within the Project Information; however, within the context of this guide, it should only be understood to mean the 'level of definition', unless otherwise stated.

LOD is defined in PAS 1192-2 as the collective term for two specific areas of information:

I level of model detail (LoD)
I level of information detail (LoI).

LoD sets out the graphical content and LoI sets out the information, or data, required to describe and maintain it at each stage of the project.

Sitting behind the dPOW are some 5,000 object templates providing guidance for defining the appropriate LoD and LoI for a number of architectural, structural and MEP (mechanical, electrical and public health) construction objects. These in turn will shape the information requirements and exchanges for each stage of the project. These objects (which become deliverables when associated with a project using the dPOW) reinforce the areas defined within the Design Responsibility Matrix, helping to align the 'what' with the 'when' and 'who' with respect to the Information Exchange requirements in response to the EIRs.

Within a dPOW, the LoD and LoI are numerically defined to reflect the work stages (these are broadly comparable to the RIBA stages, see figure 0.2), setting out guidance for what should be delivered within each Information Exchange.

A construction element can, for example, have a level of definition (LOD) of 2-2, meaning that the detail and information provided are comparable with what would normally be required at the end of the Concept Design stage (Stage 2). LoD or LoI within the LOD can also be increased at a given stage to reflect specific project design or procurement requirements. For example, a curtain wall might be developed by a specialist subcontractor at Stage 3 and the design team might therefore be required to provide an LOD of 3-3 at the end of the Concept Design, providing the specialist subcontractor with the appropriate LOD to progress the design during Stage 3.

An LOD of 2-2, as defined within the dPOW, is highlighted in figures 1.3 and 1.4.

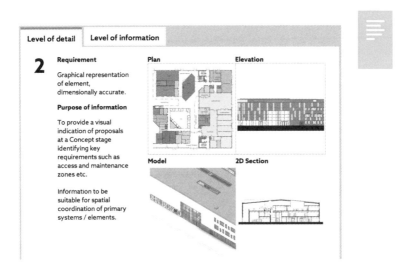

| Level of detail | Level of information |
|---|---|

**2**

**Requirement**

Graphical representation of element, dimensionally accurate.

**Purpose of information**

To provide a visual indication of proposals at a Concept stage identifying key requirements such as access and maintenance zones etc.

Information to be suitable for spatial coordination of primary systems / elements.

**Plan**

**Elevation**

**Model**

**2D Section**

*Figure 1.3  LOD: Curtain wall level of detail (LoD) at Stage 2*

## Digital Plan of Work

The Digital Plan of Work BIM Toolkit is available at https://toolkit.thenbs.com.

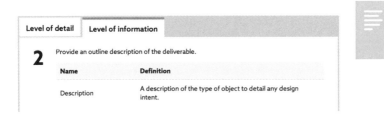

| Level of detail | Level of information |
|---|---|

**2** Provide an outline description of the deliverable.

| Name | Definition |
|---|---|
| Description | A description of the type of object to detail any design intent. |

*Figure 1.4  LOD: Curtain wall level of information (LoI) at Stage 2*

61

CIC/BIM Pro
first edition

## APPENDIX 1

### Levels of Detail and the Model Production and Delivery Table

The Levels of Detail are as follows:

LOD 1 _____

LOD 2 _____

LOD 3 _____

LOD 4 _____

LOD 5 _____

LOD 6 _____

LOD 7 _____

The Stages are as follows:

STAGE 1 _____

STAGE 2 _____

STAGE 3 _____

STAGE 4 _____

STAGE 5 _____

STAGE 6 _____

STAGE 7 _____

*This is a framework for a Model Production and Delivery Table. The parties may choose any other appropriate format and attach it to this Appendix.*

*An editable version of the BIM Protocol Appendices are provided on the BIM Task Group Website:* **www.bimtaskgroup.org**

### Specimen Model Production and Delivery Table

Showing models required at different project stages

**LOD definitions (from PAS 1192)**

1 Brief
2 Concept
3 Developed Design
4 Production
5 Installation
6 As constructed
7 In use

**Stage definitions (from APM)**

0 Strategy
1 Brief
2 Concept
3 Definition
4 Design (production information)
5 Build & Commission
6 Handover & Closeout
7 Operation and end of life

**Model Originators identified by name**

| | Drop 1 Stage 1 | | Drop 2a Stage 2 | | Drop 2b Stage 2 | | Drop 3 Stage 3 | | Drop 4 Stage 6 | |
|---|---|---|---|---|---|---|---|---|---|---|
| | Model Originator | Level of Detail | Model Originator | Level of Detail | Model Originator | Level of Detail | Model Originator | Level of Detail | Model Originator | Level of Detail |
| **Overall form and content** | | | | | | | | | | |
| Space planning | Architect | 1 | Architect | 2 | Contractor | 2 | Contractor | 3 | Contractor | 6 |
| Site and context | Architect | 1 | Architect | 2 | Contractor | 2 | Contractor | 3 | Contractor | 6 |
| Surveys | | | | | | | Contractor | 3 | | |
| External form and appearance | | | Architect | 2 | Contractor | 2 | Contractor | 3 | Contractor | 6 |
| Building and site sections | | | | | Contractor | 2 | Contractor | 3 | Contractor | 6 |
| Internal layouts | | | | | Contractor | 2 | Contractor | 3 | Contractor | 6 |
| **Design strategies** | | | | | | | | | | |
| Fire | | | Architect | 2 | Contractor | 2 | Contractor | 3 | Contractor | 6 |
| Physical security | | | Architect | 2 | Contractor | 2 | Contractor | 3 | Contractor | 6 |
| Disabled access | | | Architect | 2 | Contractor | 2 | Contractor | 3 | Contractor | 6 |
| Maintenance access | | | Architect | 2 | Contractor | 2 | Contractor | 3 | Contractor | 6 |
| BREEAM | | | | | Contractor | 2 | Contractor | 3 | Contractor | 6 |
| **Performance** | | | | | | | | | | |
| Building | Architect | 1 | Architect | 2 | Contractor | 2 | Contractor | 3 | | |
| Structural | Architect | 1 | Str Eng | 2 | Contractor | 2 | Contractor | 3 | | |
| MEP systems | Architect | 1 | MEP Eng | 2 | Contractor | 2 | Contractor | 3 | | |
| Regulation compliance analysis | | | | | | | Contractor | 3 | Contractor | 6 |
| Thermal Simulation | | | | | | | Contractor | 3 | Contractor | 6 |
| Sustainability Analysis | | | | | | | Contractor | 3 | Contractor | 6 |
| Acoustic analysis | | | | | | | Contractor | 3 | Contractor | 6 |
| 4D Programming Analysis | | | | | | | | | | |
| 5D Cost Analysis | | | | | | | | | | |
| Services Commissioning | | | | | | | Contractor | 3 | Contractor | 6 |
| **Elements, materials components** | | | | | | | | | | |
| Building | | | Architect | 2 | Contractor | 2 | Contractor | 3 | Contractor | 6 |
| Specifications | | | MEP Eng | 2 | Contractor | 2 | Contractor | 3 | Contractor | 6 |
| MEP systems | | | | | Contractor | 2 | Contractor | 3 | Contractor | 6 |
| **Construction proposals** | | | | | | | | | | |
| Phasing | | | | | | | Contractor | 3 | | |
| Site access | | | | | | | Contractor | 3 | | |
| Site set-up | | | | | | | Contractor | 3 | | |
| **Health and safety** | | | | | | | | | | |
| Design | | | | | | | Contractor | 3 | | |
| Construction | | | | | | | Contractor | 3 | | |
| Operation | | | | | | | Contractor | 3 | Contractor | 6 |

*Figure 1.5   CIC BIM Protocol Appendix 1*

The way that LOD is defined, measured and assessed is fundamental to outlining the purpose of the information produced. It underpins the content of each Information Exchange throughout the Concept Design, Developed Design and Technical Design stages as well as setting out what is required at handover for the operation and maintenance of the project during the In Use stage.

Perhaps more importantly, LOD defines the reliability of the Project Information: the extent to which it has been and can be considered by the design team at that particular stage.

The CIC *BIM Protocol's* Appendix 1, Model Production and Delivery Table (MPDT) (figure 1.5), uses an LOD system and provides guidance on how to apply the approach to each stage. While the definition notes that PAS 1192-2 sets out the basis for this LOD, the guidance does highlight that a separate standard is envisaged to set out specific requirements. It is likely that the LoD and LoI descriptions provided within the dPOW will provide the basis for this, defining the specific data content and requirements for the project. This protocol allows the BIM requirements to be appended to the project appointments so that clear responsibilities for information production are set out contractually.

Following the launch of the BIM Toolkit and the dPOW, this document and a number of other standards will be updated to reflect the LOD guidance and its specific referencing system based on project stage and content.

Other standards and protocols have been developed to help manage and define the Project Information, such as the generic and, in some cases, software-specific guidance and templates developed by AEC (UK).

### BIM standards and protocols

**AEC (UK)**
The AEC (UK) *BIM Protocol* and standards can be reviewed at
http://aecuk.wordpress.com

AEC (UK) has also produced a number of software-specific
PIM validation checklists, which outline how a model should

## BIM standards and protocols (*continued*)

be exchanged for collaboration and which are extremely useful for managing the size and appropriateness of each model file. These documents outline a best practice approach to producing and managing information. They provide a good starting point for those developing a CAD or BIM methodology, as well being a helpful reference for those who are already working with BIM standards.

### CPIc

The Construction Project Information Committee (CPIc) has also produced a suite of documents in accordance with PAS 1192-2 and aligned with the procedures outlined by the government's BIM Task Group. The CPIx protocols include pre- and post-contract BEPs, with the pre-contract plan used to address the issues raised within the government's EIRs, and the more detailed post-contract plan used to help outline the methodology that suppliers/contractors will use to deliver the project BIM.

Beta versions of the CPIx BIM strategy templates are available to download for use at www.cpic.org.uk/cpix

## Understanding roles and responsibilities

Understanding how LOD defines the extent and content of information produced during a project enables the project team to highlight who will be responsible for producing the information and what level of definition it will comprise at each specific stage. This can be defined within an MPDT, as outlined above, or through a Design Responsibility Matrix, which can subsequently assist in defining the project team's Schedules of Services.

### The Design Responsibility Matrix

The Design Responsibility Matrix is one of three tools developed in support of the RIBA Plan of Work 2013 and can be accessed via the RIBA Plan of Work Toolbox at: www.ribaplanofwork.com

## The Design Responsibility Matrix (*continued*)

The Toolbox is presented in a simple spreadsheet format, containing customisable templates for:

- the Project Roles Table
- the Design Responsibility Matrix
- the Multidisciplinary Schedules of Services.

The website and the associated RIBA publication *Assembling a Collaborative Project Team* provide detailed guidance on how to utilise the Toolbox to allocate consultants' design responsibilities, the design interface with any specialist subcontractors and the overall scope of services.

The LOD and LOI columns allow design responsibility to be defined in an 'analogue' manner (using scale and specification) or by using LOD references defined by the dPOW or elsewhere.

### How can the Toolbox be used to outline information requirements?

The dPOW sets the information required for the Design Responsibility Matrix, and once completed should export directly into scheduling software. However, the Design Responsibility Matrix can also be completed collaboratively by team members, using the RIBA Plan of Work 2013 Toolbox schedules. The Project Roles Table (figure 1.6) allows for project team members to be defined at Stage 1 of the RIBA Plan of Work. While some specifics may not be known at this stage, the general thrust and requirements of a project team can be set out.

The Design Responsibility Matrix (figure 1.7) enables the project lead to define what information is to be provided to progress the design at each stage, when it is to be provided and by whom.

The spreadsheet highlights typical building elements and systems that a project might utilise, based on the Uniclass 2 classification system. Each item can then be attributed to the disciplines outlined within the Project Roles Table using a series of simple pick lists (see Stage 2: page 79).

| | 0 Strategic Definition | 1 Preparation and Brief | 2 Concept Design | 3 Developed Design |
|---|---|---|---|---|
| Client | | | | |
| Client adviser | | | | |
| Project lead | | | | |
| Lead designer | | | | |
| Construction lead | | | | |
| Architect | | | | |
| Civil and structural engineer | | | | |
| Building services engineer | | | | |
| Cost consultant | | | | |
| Contract administrator | | | | |
| Health and safety adviser | | | | |
| Access consultant | | | | |
| Acoustic consultant | | | | |
| Archaeologist | | | | |

Figure 1.6   RIBA Plan of Work 2013 Toolbox: Project Roles Table template

The Multidisciplinary Schedules of Services (figure 1.8) outline who, following completion of the Project Roles Table and Design Responsibility Matrix, will complete specific tasks within each stage of the project. This multidisciplinary approach sets out a framework for the coordination and integration of the design information.

Aligning these ensures that the design will be progressed at the same LOD during each stage, avoiding some of the risks and errors that can occur if information is produced 'out of sync'. It also helps identify the risks of not engaging the services of some design team members during specific stages and what reworking/additional work might be required if this approach is followed.

Understanding the Project Information requirements at this stage is critical to setting out what data is exchanged and who needs to produce it. A review of these requirements at the end of this stage and Stage 2 will reaffirm the overall approach. It may also start to highlight specific procurement approaches, which may then impact what information is to be provided, when and by whom.

| Classification | Aspect of design — Title | 2 - Concept Design — Design team | | | 3 - Developed Design — Design team | | |
|---|---|---|---|---|---|---|---|
| | | Design responsibility | Level of detail | Level of information | Design responsibility | Level of detail | Level of information |
| 15-05 | Substructure | | | | | | |
| 15-05-65 | Piling | | | | | | |
| 15-65-75 | Insitu concrete frame | | | | | | |
| 15-65-75 | Post tensioned concrete frame | | | | | | |
| 15-65-75 | Precast concrete frame | | | | | | |
| 15-65-75 | Steel frame including secondary steel | | | | | | |
| 20-10-20 | Suspended ceilings | | | | | | |
| 20-15-05 | Hard landscaping | | | | | | |
| 20-25-75 | Roof lights | | | | | | |
| 20-50-30 | Flat roof systems | | | | | | |
| 20-50-50 | Metal sheet roof systems | | | | | | |
| 20-55 | Carpets and other floor finishes | | | | | | |
| 20-55-15 | Screeds | | | | | | |
| 20-55-35 | Internal floor tiling | | | | | | |
| 20-55-70 | Raised access floors | | | | | | |
| 20-55-95 | Timber flooring | | | | | | |

*Figure 1.7 RIBA Plan of Work 2013 Toolbox: Design Responsibility Matrix (aligned with dPOW)*

67

## Multidisciplinary Schedule of Services

### 1 – Preparation and Brief

| Project role | Party | Tasks to be undertaken |
|---|---|---|
| All roles | | Provide information for and contribute to contents of Project Execution Plan as required |
| Client and/or client adviser | 0 | Contribute to development of Initial Project Brief including Project Objectives, Quality Objectives, Project Outcomes, Sustainability Aspirations, Project Budget and other parameters or constraints |
| Project lead | 0 | Develop Initial Project Brief with project team including Project Objectives, Quality Objectives, Project Outcomes, Sustainability Aspirations, Project Budget and other parameters or constraints |
| | | Collate comments and facilitate workshops as required to develop Initial Project Brief |
| | | Prepare Project Roles Table and Contractual Tree and continue assembling and appointing project team members |
| | | Prepare Schedule of Services and develop Design Responsibility Matrix including Information Exchanges with lead designer |

Figure 1.8    RIBA Plan of Work 2013 Toolbox: Multidisciplinary Schedule of Services

## Chapter summary    1

The Preparation and Brief stage defines the Initial Project Brief. This document will be developed and supported through a number of Feasibility Studies based on the collated and reviewed Site Information, the requirements for assembling the project team, the project scope and the associated design responsibilities.

Through the development the Technology and Communication Strategies, defined within the Project Execution Plan, the project can define protocols and standards collaboratively to manage the information production and workflow, either in 2D/3D CAD or within a BIM environment. This will allow the structure and accuracy of data collected and produced to be standardised and verified in a usable format so that the key information can be utilised within the Feasibility Studies, the Initial Project Brief and throughout the future design stages, minimising any reworking and inefficiencies created through incompatible processes, scopes and approaches.

It is important to ensure that, as it develops, the information production and management is controlled through the Project Strategies and that the disciplines responsible for its completion in this and any future stages are aware of what should be produced, for what purpose and to what level of detail. Outlining these requirements early within the project will allow each discipline to programme resources appropriately. This will ensure that all areas of the design are suitably tested and explored, improving both the efficiency and quality of the overall process and project.

The Stage 1 Information Exchange is:

| Initial Project Brief.

Additional Information reviewed might include:

| Site Information
| Feasibility Studies
| Project Budget
| Sustainability Aspirations
| Risk Assessments
| appointment documentation
| Digital Plan of Work
| Design Responsibility Matrix
| Project Strategies.

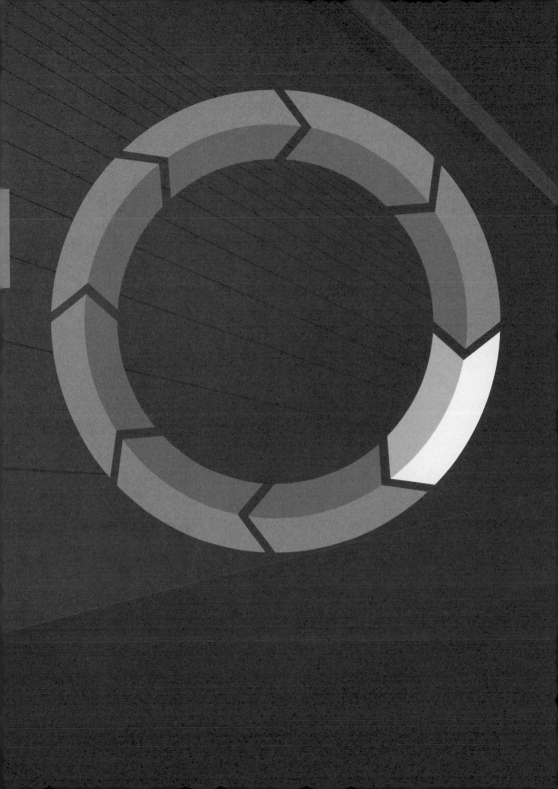

Stage 2

# Concept Design

# Chapter overview

The information developed during Stage 2 informs the Final Project Brief and the Concept Design. The amount of information completed at this stage will depend on the complexity, size and nature of the project, but the stage information and briefing document should endeavour to communicate all the strategic design principles.

**The key coverage in this chapter is as follows:**

Understanding the impacts of the variable task bars

Developing the Final Project Brief

What are the Information Exchanges at the end of Stage 2?

Defining information protocols and standards

Developing the Concept Design

# Introduction

As the design approach narrows toward a single option, the project team can progress a number of strategies to inform the developing design. These may suggest what the best procurement strategy might be and how it might impact on the way Project Information is produced and who is to produce it. The planning strategy will be considered within the context of the emerging Concept Design and developing Construction Strategy, outlining when might be appropriate for the submission of a full planning application or when a pre-application consultation might occur.

The development of these strategies will directly impact the content, programme and level of definition (LOD) of the information to be produced and exchanged throughout this and subsequent stages. Reviewing these requirements against the Design Responsibility Matrix will ensure that any gaps in the information and scope are removed and that the appropriate design team members are appointed and are working collaboratively.

With the design approach and strategy developing clarity at this stage, it is important to minimise any errors and inconsistencies that occur when systems are developed ad hoc and in isolation. This is achieved by ensuring that any information produced by the design team is structured and managed in accordance with the protocols and standards established during the previous stage.

## What are the Core Objectives of this stage?

The Core Objectives of the RIBA Plan of Work 2013 at Stage 2 are:

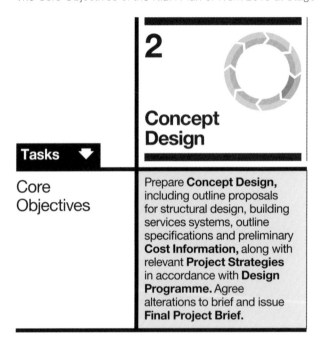

| | 2    Concept Design |
| --- | --- |
| **Tasks** ▼ | |
| Core Objectives | Prepare **Concept Design,** including outline proposals for structural design, building services systems, outline specifications and preliminary **Cost Information,** along with relevant **Project Strategies** in accordance with **Design Programme.** Agree alterations to brief and issue **Final Project Brief.** |

Stage 2 will commence by preparing Concept Design proposals using the Initial Project Brief and the Site Information. As Stage 2 progresses the brief will evolve alongside the Concept Design and will be issued as part of the Information Exchange at the end of the stage. The key Project Strategies and approaches will progress accordingly, aligned with the design solution, and should be reviewed and documented within the Project Execution Plan at the end of the stage and summarised within the Final Project Brief.

# Understanding the impacts of the variable task bars

Programming, procurement and the submission of a planning application are all project activities that can be difficult to set out definitively at the outset of the design process. Each tends to influence the other and, more importantly, the information requirements and the LOD within each particular stage. The RIBA Plan of Work 2013, in its digital format, allows for a project-specific Plan of Work to be developed against each of these three activities using a specific task bar and pull-down list. Selecting an option for one activity can suggest the direction that another might take, and how the subsequent tasks, objectives and Information Exchanges might be affected. The approach to these activities should be reviewed as early as possible, and although it might not be finally resolved until a later stage, an understanding of its impacts can help to inform the Design Responsibility Matrix, completed during Stage 1, and also the Schedules of Services of the core project team members.

### Can a planning application be made at this stage?

The project team may decide to submit a planning application at the end of this stage, prior to the RIBA-recommended stage for a detailed planning application – Stage 3. This decision will need to be reviewed as early as possible as it will impact the design development and content of the Project Information for this and subsequent stages.

In order to ensure that a Stage 2 planning application is sufficiently coordinated and avoids risks associated with future design changes, the information will need to be developed collaboratively, with the appropriate input from the design team. If the design strategies are not sufficiently progressed prior to an application then there will be a high likelihood that the design will have to be reworked and a revised or amended application submitted. Locating the Concept Design on site without having a full understanding of how the Sustainability Strategy might be impacted, or fixing the height of the design without having an understanding of the servicing strategy, highlights the type of risks associated with developing the design without this collaborative input. If the design team is not fully appointed at this stage then these and any other associated risks should be clearly outlined to the client prior to progressing the design.

### The planning process

For detailed advice on the planning process, refer to the *RIBA Plan of Work 2013 Guide: Town Planning*, published as part of this series, in conjunction with the UK government's website: www.planningportal.gov.uk

### How does procurement impact the Information Exchanges?

Procurement will tend to follow a strategy that best delivers the client's objectives, balancing any associated risks against the required Project Outcomes. More often than not the key areas that define the procurement strategy will conflict with each other. These must therefore be carefully reviewed to ensure the right balance is achieved to support and deliver the client's aspirations.

Typically, time, cost and quality represent the main procurement drivers: in most cases, two out of these three will have more importance to the overall project success. Certainty of quality and cost, quality and time or, more commonly, cost and time in association with the Project Objectives outlined within the Initial Project Brief will set out which of the criteria are the most important and which may offer the greatest risks.

The RIBA Plan of Work Procurement task bar currently outlines five principal procurement routes, the majority of which are driven by time and cost parameters, although in most cases quality will be managed through the detail provided within the contract information.

The five procurement routes have specific Information Exchange requirements, which are typically as follows:

l  Traditional contract
All design information is provided at the end of Stage 4.

l  One-stage design and build contract
Employer's Requirements are provided for tender at the end of Stage 3 and developed as contract information thereafter.

  |  Two-stage design and build contract
Employer's Requirements are provided at the end of Stage 4 as contract information. Tender information will typically be provided after Stage 2 or 3.

  |  Management contract
All design information is finalised in line with the procurement strategy.

  |  Contractor-led contract
All design information is developed by the contractor in line with the procurement strategy.

The choice of procurement route will significantly impact the purpose of the Project Information at each stage, with the Design and Project Programmes, the management of change and costs and the LOD all being dependent on when information is required and from whom. It is important to ensure that both the procurement strategy and the Construction Strategy are reviewed as early as possible by the whole project team to ensure the information required is produced at the appropriate stage and that the client's objectives are maintained.

The traditional contract route is typically the one most commonly used. As the design is fully developed before tender, the client has greater certainty with regard to design quality and cost. This route represents the only form of procurement where the design team alone produces the contract information. There are variations within this route that allow for elements of contractor design, but for the majority of projects traditional procurement offers no pre-contract buildability input from the contractor or specialist subcontractors during the development of the design.

While the traditional route does offer some cost and quality certainty, it is also the most adversarial approach and does not lend itself to collaborative working in the later stages of the project design (Stages 3 and 4).

Developing information in collaboration with the full project team can facilitate a more integrated design and construction process, especially with the earlier input of specialist subcontractors and suppliers. This approach may have an impact on the design costs, but it should ensure fewer coordination issues develop as the project progresses. Design and build and contractor-led procurement routes can provide this type of collaborative approach, although the quality of the project will be governed by the accuracy and content of the information developed to define the Employer's Requirements.

The production of information within a BIM environment requires a cooperative approach from the whole project team. While different procurement routes can require the design disciplines to be restructured during the course of the project, the collaborative nature of the information is key to avoiding coordination issues, inefficiencies, clashes and waste during the construction stages. The client should consider the broad principles of collaborative practice as early as possible, so that the appropriate procurement route can be chosen.

### How can the Design Responsibility Matrix inform the design?

The Design Responsibility Matrix (DRM) will be generated in Stage 1 to inform the scope and composition of the design team. Much of the information outlined should be revisited in subsequent stages to reflect any adjustments that the emerging design may require.

**2 - Concept Design**

| Aspect of design | | Design team | | |
|---|---|---|---|---|
| Classification | Title | Design responsibility | Level of detail | Level of information |
| 15-05 | Substructure | Civil and structural engineer | 2 | |
| 15-05-65 | Piling | | | [Not decided] |
| 15-65-75 | Insitu concrete frame | | | [Not required] 1 |
| 15-65-75 | Post tensioned concrete frame | | | 2 3 |
| 15-65-75 | Precast concrete frame | | | 4 5 |
| 15-65-75 | Steel frame including secondary steel | | | 6 7 |
| 20-10-20 | Suspended ceilings | | | |
| 20-15-05 | Hard landscaping | | | |
| 20-25-75 | Roof lights | | | |
| 20-50-30 | Flat roof systems | | | |
| 20-50-50 | Metal sheet roof systems | | | |
| 20-55 | Carpets and other floor finishes | | | |
| 20-55-15 | Screeds | | | |

*Figure 2.1   DRM as downloaded*

The DRM will outline the design information required from each design team member in relation to the stage and set out individual design responsibilities and the LOD required. (The release of the dPOW BIM Toolkit redefines the way LOD is specified and it is expected that the DRM template will be revised to utilise the same system. In the short term, this can be amended manually if parity is required.) Formalising what will be produced, by whom and when is an extremely important step in the development of the project, defining the scope and the format of the information to be exchanged and reviewed, in the context of the Project Programme.

The basic DRM template as downloaded from the RIBA Plan of Work Toolbox (www.ribaplanofwork.com) (figure 2.1) defines specific areas of information by referencing their Uniclass code. While these can be applied to either a BIM or CAD approach, it may be more appropriate to amend the pick list content to suit a less detailed approach at this stage or to use the high-level Uniclass definitions (see page 83).

Figure 2.2 shows the 'Aspect of design' headings amended, merging the 'Classification' and 'Title' columns into a single column, titled 'Project Information', which allows a 'Drawing and task' list to be viewed in a drop-down pick list generated to help populate each field. This list could also comprise the stage information checklists highlighted within this guide to help outline scope, responsibility and programme.

Also, this version shows 'analogue' deliverables rather than LOD. A version generated from the dPOW Toolkit would always use LOD, with 'analogue' deliverables defined separately.

Adding information in drop-down lists and customising the DRM is not a prerequisite, but it may make it easier for the project team to produce a coordinated matrix during Stage 1, especially when the specific details of the design are not defined.

### Amending pick lists

The integral Help menu in most spreadsheet software will outline how to produce and manage drop-down lists within the application, which will be helpful for defining and creating pick lists within the DRM. Alternatively, additional columns can be added and the information added manually by each member of the project team.

| 2 - Concept Design | | | |
|---|---|---|---|
| **Aspect of design** | **Design team** | | |
| **Project information** | **Design responsibility** | **Level of design** | **Information exchange** |
| Ground floor plan GA | Architect | Outline | 1:200 |
| First floor plan GA | Architect | Outline | 1:200 |
| Second floor plan GA | Architect | Outline | 1:200 |
| Typical floor plan GA | Architect | Outline | 1:200 |
| Roof plan GA | | | |
| Section GA | | | |
| Elevation GA | | | |
| Grid File | Structural Engineer | Outline | 1:200 |
| | | | |
| Specification: Outline<br>Specification: Performance<br>Specification: NBS<br>Specification: RDS<br>Door schedule<br>Window schedule<br>Finishes schedule<br>Area Schedule | | | |

**Note:** The 'Level of design' and 'Information exchange' headings have now been adjusted to 'Level of detail' and 'Level of information', in line with the dPOW (see figure 2.3).

*Figure 2.2   Early approach to the DRM (using 'analogue' deliverables)*

The DRM is a flexible document and a number of different approaches can be used. For example, in figure 2.3 the 'Aspect of design' column outlines the specific CAD model files to be produced. This is a less detailed method than setting out the published Information Exchanges, but it may be a more helpful approach as it allows the project team to quickly define the scope and programme.

Figure 2.4 highlights the latest Uniclass 2 definitions within a pick list, ordered so that each discipline can review and assemble the information it will develop in relation to the scope of the project. This approach can be more suited to a BIM environment, where elements and objects can be defined and placed using the same Uniclass codes and references.

| 2 - Concept Design | | | |
|---|---|---|---|

| Aspect of design | Design team | | |
|---|---|---|---|
| **Project Information** | **Design responsibility** | **Level of detail** | **Level of information** |
| Site Boundary Model | Client | 2 | 2 |
| External works Model | Civil engineer | 2 | 2 |
| Ground Floor Model | Architect | 2 | 2 |
| First Floor Model | Architect | 2 | 2 |
| Grid Model | Lead designer | 2 | 2 |
| Ground Floor Structural Model | Structural engineer | 2 | 2 |
| Ground Floor Mechanical Model | Building services engineer | 2 | 2 |
| Ground Floor Electrical Model | Building services engineer | 2 | 2 |
| Section Model | Architect | 2 | 2 |
| Structural Section Model | Civil engineer | 2 | 2 |
| | | | |

Roof Model
Section Model
Elevation Model
Fire Strategy Model
✓ Grid Model
Ground Floor Structural Model

*Figure 2.3    dPOW approach to the DRM (using LoD and LoI)*

## What is Uniclass?

Uniclass is the predominant classification tool for the construction industry, having superseded the previous CI/SfB system (although many architects still use CI/SfB for numbering drawings and organising information). Uniclass organises and classifies all forms of construction information, including building components, spaces, infrastructure and management systems, from the initial concept through to project handover, and to the operation and maintenance of the project.

## 2 - Concept Design

| Aspect of design | | Design team | | |
|---|---|---|---|---|
| Classification | Title | Design responsibility | Level of detail | Level of information |
| Ee_15_50_50 | Building Surveys | Client | 2 | 5 |
| Ee_25_20 | External Walls | Architect | 2 | 2 |
| Ee_25_25 | Internal Walls | Architect | 2 | |
| Ee_30_10 | Roofs | Architect | 2 | [Not decided] [Not required] |
| Ee_30_20_05 | Lowest Floors Substructure | Structural engineer | 2 | 1 |
| Ee_30_20_10 | Lowest Floors Structure | Structural engineer | 2 | 2 3 |
| Ee_30_40 | Upper Floors | | | 4 |
| Ee_30_40_10 | Upper Floors Structure | | | 5 |
| Ee_30_40_10 | Upper Floors Structure | | | |
| Ee_35_10 | On-Ground Stairs | | | |
| Ee_60_50_40 | Building Combined Heating And Cooling | Building services engineer | Outline | |
| Ee_60_50_44 | Building Cooling | | | |
| Ee_60_50_48 | Building Heating | | | |
| Ee | Elements | | | |

*Figure 2.4   Completing the DRM*

Uniclass 2 has recently been developed to align all the original classification tables, using the same numerical system to facilitate cross-referencing, to ensure that each code is relative throughout the project's development.

### Digital Plan of Work definitions

The Uniclass classifications associated with the Digital Plan of Work's 5,000 objects will reflect the latest developments in the system and, if these lists are produced within the NBS BIM Toolkit, the information can be exported directly into the DRM. This data can be used to create a project-specific pick list of objects, which can then be aligned with the appropriate designer and stage requirements.

The basis for Uniclass 2 has now been finalised and is available online at the CPlc's website (see box below).

Currently, the Uniclass 2 tables comprise:

**Co**    Complexes
**En**    Entities
**Ac**    Activities
**Sp**    Spaces
**EF**    Entities by form
**Ee**    Elements
**Ss**    Systems
**Pr**    Products
**Zz**    CAD
**PP**    Project phases.

The Uniclass element naming system ensures that an element always retains its original identity and meaning within the project process, allowing information to be created and reused by many different organisations throughout the life cycle of the project.

### Uniclass classification system

CPlc – the Construction Project Information Committee – currently governs the definitions and classification tables for the Uniclass system, which can be downloaded from its website:
www.cpic.org.uk

Uniclass 2 is also available to view in full at:
www.cpic.org.uk/uniclass2

For a detailed summary of how Uniclass 2 has been developed and further links to explanations of how the specific tables and specifications work, refer to: www.thenbs.com/uniclass/

The dPOW BIM Toolkit is supported by Uniclass 2015. This system builds upon the Uniclass and Uniclass 2 definitions to provide a comprehensive classification system for use by the entire industry, and for all stages of the project. For more details on Uniclass 2015 refer to: https://toolkit.thenbs.com/articles/classification

As outlined earlier, the Uniclass format can be used to set out design responsibilities using the less detailed 'parent' levels, such as using Ee_25_20 External Walls to represent the envelope. This can then be refined during a later stage to reflect a system category, such as Ss_25_13_50 Masonry Wall Systems, when the design is more resolved. The lead designer may update the DRM to reflect this more detailed approach when the information reaches a specific stage or, subject to the procurement route, when the contractor takes responsibility for the development of the information.

Adding further detail to reflect the developing design might also highlight the need for a change in design responsibility. This should be recorded appropriately as it may have an impact on individual discipline scopes, design costs and the Project Programme.

While CAD and BIM systems are still developing, we are not yet at the point where the digital model file alone can satisfy all the requirements of a construction project: the more traditional plans, sections and elevation views are still required throughout a project, issued to scale, within a drawing sheet file. These associated outputs and tasks will be required during and at the end of each stage and the DRM should ensure that their production is captured within each discipline's scope, whether as the main output requirement, an additional list or outlined within the 'Notes' column of the DRM.

### Designing with costs

The construction cost estimate developed during the Concept Design stage will typically be based on a £/m$^2$ rate measured from any drawings and models produced by the design team. The level of certainty for the costs will depend on the resolution of the design and how well it reflects the requirements of the brief.

Often, designs produced at this early stage may have only benefited from generic advice regarding structural, servicing or environmental strategies. Progressing the design with a limited design team can result in abortive and costly reworking, distorting and possibly increasing the overall project costs. However, using a collaborative project team at this or an earlier stage will enable the Project Information to be fully developed and, as a result, offer greater certainty of costs and reduce any associated financial risks.

The accuracy of the Cost Information at this stage will depend on the level of definition and the detail and information included. While it will typically be quite generic at Stage 2, it is possible that some detailed information can be contained in the model files, especially within BIM objects and components. This information could be measured and costed, but the project team should ensure it is only utilised if the level of definition included is that intended by the originating discipline.

While this data might not necessarily be used to assess the efficiencies of the design, if included it can offer an insight into how the cost plan might develop and may ultimately assist the progression of the design in later stages. The Project Execution Plan and Technology Strategy will set out the required LOD for this stage and all cost data should be reported against these requirements.

## Developing the Final Project Brief

The Final Project Brief represents the basis from which the project design will be developed. It is the last iteration of the briefing process and summarises the principles of the Initial Project Brief developed and finalised over the course of Stage 2. The brief will highlight the agreed concepts, performances and parameters and will be approved and signed off in association with the Concept Design at the end of the stage.

### The Final Project Brief – information checklist

Typically, the Information developed within the Final Project Brief will include:

General
- the client's objectives, requirements and established priorities and criteria
- the quality standards
- the project environmental policy
- the context, including site history, topography and geology
- the site boundary
- studies previously commissioned
- plans for future expansion
- other likely stakeholders' requirements
- an assessment of known project risks

## The Final Project Brief – information checklist
**(continued)**

Procedural

- key project milestones and programme
- project roles and scopes, highlighting disciplines required to carry out the design
- strategies and requirements to be developed within the Project Execution Plan

Functional

- accommodation schedule
- space standards
- spatial relationships
- access requirements
- flexibility to accommodate future reorganisation
- allowance for future expansion or extension
- lifespan for structure, elements and installations
- Operational and Maintenance Strategy
- special considerations (eg security, acoustics)
- servicing options

Environmental constraints

- any specific site constraints
- summary of Site Information

Statutory

- summary of planning constraints
- summary of planning approach
- summary of legislative constraints

Financial

- the Project Budget
- benchmarking and approximate $£/m^2$
- operating expenditure, costs in use and other whole-life costs
- review of the likely procurement strategy

Change tracker

- a summary of revisions and inclusions to the brief requirements and strategy.

As with the Initial Project Brief, this document may include significant amounts of detailed information and so the use of an executive summary should be considered to draw out the key requirements and objectives developed during the Concept Design stage. This type of summary will allow some clients and stakeholders to gain a concise overview of the key objectives and strategies without having to review the full content.

The Final Project Brief should be structured to align with the documents that have been produced previously, allowing the information to be compared and transferred as required without significant rewriting or reworking.

## What are the Information Exchanges at the end of Stage 2?

The Information Exchanges for Stage 2 comprise:

I the Concept Design
I the Final Project Brief
I the Project Strategies
I Cost Information.

This stage also has an equivalent exchange within the UK government's Digital Plan of Work (dPOW):

I UK Government Information Exchange: Data Drop 2.

### UK Government Information Exchanges

The Data Drops in the dPOW are outlined on the BIM Task Group's guide *COBie Data Drops: Structure, uses and examples* (March 2012), which is available on its website: www.bimtaskgroup.org/cobie-data-drops/

The dPOW is structured to prefer a contractor-led approach to project delivery, therefore Data Drop 2 defines the level of definition required to tender the design, with information based on a high certainty or accuracy. The data will be broadly consistent with a Concept Design developed to provide Employer's Requirements and should include sufficient detail to define the design and allow a main contractor to be selected competitively.

The information production at this stage is managed over two distinct phases: Data Drops 2a and 2b.

I Data Drop 2a identifies requirements for a Level 2 (BIM) federated model information to be delivered by the client sufficient to tender the design.
I Data Drop 2b identifies requirements for a Level 2 (BIM) federated model information to be delivered by the contractor to reflect their price.

The two proposals will be compared and any differences or changes will identify any Contractor's Proposals and any areas of non-compliance to be resolved prior to progressing to the next stage.

In addition to the Data Drop, a number of plain language questions have been set out to inform the design at this stage sufficiently for it to progress to Stage 3. These cover areas such as:

I BIM standards and protocols
I the LOD as defined in the Model Production and Delivery Table
I surveys and context
I the concept design
I design performance
I structural concepts
I MEP (mechanical, electrical and public health) concepts
I coordination zones
I statutory compliance.

## What are the Key Support Tasks and Project Strategies?

The Suggested Key Support Tasks outlined at this stage include preparing or reviewing the:

I Sustainability Strategy
I Maintenance and Operational Strategy
I Handover Strategy
I Risk Assessments
I Construction Strategy
I Health and Safety Strategy

I Project Execution Plan, including Technology and Communication Strategies
I Research and Development.

As the Concept Design develops and the Final Project Brief is completed, the Project Execution Plan and associated Project Strategies set out in Stage 1 should be reviewed and updated where appropriate.

The project's buildability and approach, such as offsite or component-driven construction, will be explored in the Construction Strategy, which will also be aligned with the procurement strategy and developed to reflect the client's requirements for delivering the final project. These strategies will highlight when the main contractor will be involved in the development of Project Information and what impact delivery and buildability will have on the level of definition produced within this and subsequent stages.

The Health and Safety Strategy should also be developed with the Construction Strategy and should highlight any significant design risks that arise during the review of the Site Information, constraints and the early design development. The designer's Risk Assessment should also identify where risks might be mitigated by the Concept Design and how risks will be managed during construction or the operation of the facility.

## Defining information protocols and standards

Controlling the way information is created and understanding how it can be used by all disciplines within the project team, whether produced within a CAD or BIM environment, will require the rigorous application of standards and protocols. Those set out within BS 1192:2007 and developed in PAS 1192-2:2013 and associated documents should be used as a basic starting point for managing the design.

### What standards help to manage the Project Information?

BS 1192 highlights eight principles for the implementation of a standard method and procedure (SMP):

1. Roles, responsibilities and authorities
2. Common data environment (CDE)
3. Document management or electronic document management (DM/EDM) system

4. File naming convention
5. Origin and setting out
6. Drawing sheet templates
7. Layer standard
8. Annotation.

File naming, drawing templates, layer standards and annotations all relate to the way information is exchanged and controlled and represent the minimum required to support the implementation of a CDE and DM/EDM system.

### What is a common data environment?

A CDE is often described as a single location for managing and storing information, typically on a server or web-based system, in support of a collaborative workflow. While this definition reflects the use of a CDE, it should, however, be considered as a collaborative *process*, used to structure the development and exchange of design and asset information produced by all members of the project team and all other project stakeholders.

Managing information efficiently through a productive working method will save time and avoid the need for reworking due to information inaccuracies. Using the processes defined by a CDE supported by the SMP set out in BS 1192 will ensure the project develops validated information at each stage. These collaborative systems also ensure that a properly coordinated set of information is iteratively progressed from the outset of the project, in accordance with its stated purpose, such as information, design coordination or construction. Figure 2.5 outlines the framework of the CDE processes set out within BS 1192.

### Standards and protocols

BS 1192 outlines the process and procedures defined by the SMP and required by a CDE. Further detail on the process and application can be found in *Building Information Management: A standard framework and guide to BS 1192*, by Mervyn Richards (British Standards Institution, 2010).

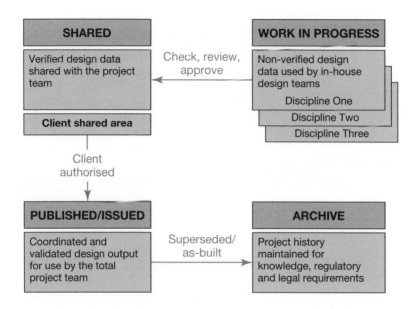

*Figure 2.5    Principles of a CDE*

Both BS 1192 and PAS 1192-2 highlight that a CDE can be applied to all information formats, including:

I    the development and coordination of 2D model files
I    the development and coordination of 3D model files
I    the development and coordination of 2D files from 3D models
I    production of 2D drawings from CAD software
I    the development, coordination and management of the specification and costing documentation
I    collection, management and distribution of documentation
I    management of all project-related information, including information from a BIM model
I    management of protocols for the delivery of a fully coordinated and integrated building information model (iBIM).

The process allows for project information to be produced, maintained and developed either locally by a specific discipline or, alternatively, through a web-based site. Web-based solutions might be necessary for managing the workflows of larger projects; however, the principles of the CDE are

## Using a CDE

Managing small projects using a CDE will focus primarily on the process of information management, rather than how and where that information is stored. Small projects should endeavour to follow the structure of the CDE when exchanging information, ensuring the status is either 'work in progress' (WIP), 'shared', 'published' or 'archived', and that these uses are understood within the team. Primary information can then be stored locally as necessary by each discipline, with the most recent files and data shared for collaboration.

applicable to projects of all sizes and scales and can be implemented and operated irrespective of whether there is a centralised data location.

## Why use standard methods and procedures?

Using the BS 1192 SMP in the production of Project Information will ensure that many of the issues attributed to poor coordination and management, such as inaccuracy, errors and inefficiency as a result of poorly produced information, do not occur and that the quality of the design information and productivity improves accordingly.

The principles used within the project should be defined within the Technology Strategy developed in Stage 1 and reviewed and agreed by all project team members at each stage. This will ensure that if additional design team members are appointed or existing ones replaced, the processes and procedures are clear, deliverable and practical to all. The efficiencies gained in the production of information and the quality of that information may be compromised if stage reviews are not followed.

One of the key standards for facilitating collaboration outlined within the SMP is that for origin and setting out. This is a fundamental process in both CAD and BIM approaches, although its impact may be more significant within a Project Information Model (PIM) as the information is positioned in three dimensions.

All digitally produced design information should be geospatially located in accordance with the real world coordinates of the site.

### Geospatial locations

BS 1192 describes a geospatial location as a: 'system in three dimensions (decimal degrees latitude, longitude and elevation in metres) and a plan orientation (decimal degrees clockwise rotation from north)'.

Often, at the outset of a project, appropriate survey data does not exist and so Ordinance Survey coordinates are used to locate an area of the site from which a setting-out point can be agreed. Ordinance Survey coordinates can cause dimensional issues in some software, therefore these should be replaced by an accurately surveyed point as soon as possible to ensure that all information can be set out relative to the site coordinates. A building or project grid/grids can then be produced with reference to a relative site grid or setting-out point.

Agreeing a project setting-out point relative to the site and the design will ensure that all disciplines can, in their native software, align information so that when exchanged it will be imported in the correct location. This process should be tested between all disciplines as significant time can be lost if files do not overlay correctly. Disciplines should never relocate the file of another discipline to suit their own or the project grid. This adds unnecessary risk to the coordination process as the file may be relocated incorrectly, not only in location but also in rotation and, in the case of a BIM model, at the wrong height.

Exporting information with the grid file will allow each discipline to test that they can import the information correctly. An insertion point should always be included within the transferred file for checking purposes.

### Locating the model file data

The file origin (0,0,0) should always be outside the area of the project to avoid negative values in the model files and, in a BIM model, it should always be close to the location of the project. If the model is too far away from the 'centre' of the design file then errors can occur in dimensioning, drawing and exporting. It is important to follow the software's process for assigning a global origin point in relation to 'real world' coordinates for the project.

## Standards and protocols

These processes are outlined in more detail in *Building Information Management: A standard framework and guide to BS 1192*, by Mervyn Richards (British Standards Institution, 2010), and also within the AEC (UK) BIM Protocol and the AEC (UK) standards for particular software, which are particularly useful for BIM setting out.

The AEC (UK) documents include both CAD and BIM standards and protocols, including:

- AEC (UK) CAD Standards For Layer Naming v3.0
- AEC (UK) Model File Naming Handbook v2.4
- AEC (UK) Drawing Management Handbook v2.4
- AEC (UK) BIM Protocol v2.0 (Main document)
- AEC (UK) BIM Protocol – BIM Execution Plan v2.0
- AEC (UK) BIM Protocol – Model Matrix v2.0
- AEC (UK) BIM Protocol Model Validation Checklists for Autodesk, Bentley and Graphisoft systems.

AEC (UK) standards can be downloaded from:
http://aecuk.wordpress.com

All the standards should be read in conjunction with BS 1192 and PAS 1192-2.

## Developing the Concept Design

The information delivered at the end of Stage 2 will be developed collaboratively from the concepts and strategies of each design team member and should reflect the stage requirements set out within the Design Responsibility Matrix. With a CAD approach, this will mean documenting the size and extent of the spaces and their adjacencies, with the major structural and servicing zones identified in principle at floor and ceiling interfaces so that the overall size and height of the design is finalised. In addition, the design will set out areas for servicing and incoming services, the outline Sustainability Aspirations and relationships between context and materials. The design will also identify areas that may need further input from specialist designers/subcontractors. This will allow the responsibility for the interfaces to be understood so that they can be coordinated appropriately within the later stages.

If a BIM approach is used, the same interfaces and zonal requirements will be highlighted, but primarily in three dimensions, allowing the design to be understood visually by the whole project team as it develops. This ensures that coordination issues and constraints inform the design at this stage as opposed to defining it.

The Initial Project Brief should be reviewed as the design progresses, and any feedback received from the client and other stakeholders on the Stage 1 information incorporated. Any additional survey information should be identified, completed and integrated into the Concept Design before the stage is complete.

### What information defines the Concept Design?

For the foreseeable future, and irrespective of the process through which the design is developed, the information needed to review and record progress will inevitably be issued as some form of drawing. However, while general arrangement drawings (GAs) might typically be used to describe the proposal at this stage, it is likely that 3D models, BIM models, views, sketches and visualisations will also now be used to communicate the design intent and compliance with the brief.

Changes in the procurement of design services can occur at the end of this and subsequent stages and the standard, format and content of the Project Information will need to be developed to facilitate the seamless progression of the project. If the information is not aligned with the strategies within the Project Execution Plan or is not produced rigorously, then the revised design team can lose focus, with time and cost consequences that may impact the required outputs or intent during later stages.

A Stage 2 report summarising the Final Project Brief, Project Execution Plan and Concept Design information will ensure that the design intent is documented and communicated appropriately at the end of the stage.

## Stage 2 report – information checklist

### Introduction

– Summary of Final Project Brief
– Summary of Project Execution Plan, including:
  ○ general Project Information
  ○ project team key contacts
  ○ key stakeholder requirements
– Brief tracker and information required schedule, including:
  ○ summary of changes to the brief
  ○ summary of information required to develop the design within Stage 3
  ○ summary of unresolved information that may impact Stage 3 design

### Site Information

– Summary of survey information

### Architectural

– Introduction
– History and conservation
– Summary of design requirements
– Design approach
– Site analysis and constraints
– Schedule of accommodation
– Accommodation adjacencies
– Functional layouts
– Concept Design information
– Circulation/operation strategy
– Outline specification and finishes
– Access and egress strategy (might be a separate section, depending on type, size etc.)
– Phasing/demolition
– Town planning analysis and implications
– Summary of discussions with statutory authorities, including local planning authority
– CDM regulations and responsibilities
– Access strategy for cleaning and maintenance

### Structural, civil and geotechnical engineering

– Introduction
– Desk study/site analysis
– Report on existing facilities and engineering systems, if applicable
– Design criteria/aims and objectives
– Outline design information
– Outline specification

## Stage 2 report – information checklist (*continued*)

- Underground drainage
- Sustainability
- Cleaning/maintenance
- Environmental
- Demolition
- Civils summary

### Mechanical and electrical engineering

- Introduction
- Desk study/site analysis (existing gas, water, electrical, mechanical etc. services)
- Report on existing facilities and engineering systems, if applicable
- Design criteria/aims and objectives
- Design concept/proposed services strategy
- Outline design information
- Outline specification

### Environmental strategy

- Basis of design study
- Siting options and climatic influences
- Massing models
- Relationships to site context
- Outline design information
- Outline specification
- Transport assessment

### Specialist design strategies

- Reports produced by any specialist consultants, including:
  - acoustic or fire engineering
  - arboriculture and ecology
  - rights to light
  - human factors
  - vertical transportation

### Construction Strategy

- Buildability issues
- Access and logistics

### Cost Information

- Stage 2 construction cost estimate
- Preliminary cost estimate

## Stage 2 report – information checklist (*continued*)

- Concept schedule of materials and finishes.
- Value engineering proposals

### Programme

- Project Programme
- Phasing
- Design Programme
- Programme improvement proposals

### Procurement

- Proposed Procurement Strategy
- *OJEU* notices (if applicable)

### Risk

- Summary
- Risk Register (including preliminary architectural, structural, M&E and cost risks)
- Highlight 'special' project risks

### Health and Safety Strategy

- Summary
- Integration of health and safety issues and risk into the design
- 'Significant' or unusual buildability and health and safety issues
- Priority health and safety issues

### Operational and Maintenance Strategy

- Summary
- Approach and requirements

### Key performance indicators

- Recommended KPI targets for the project

### Way forward

- Outline of elements not covered in Concept Design
- Way forward

### Appendices

- Summary of client and stakeholder project reviews
- Architectural strategy, can typically include (depending on type and size of the project):

## Stage 2 report – information checklist (*continued*)

- ○ masterplans
- ○ parameter plans
- ○ areas
- Conceptual drawings, including
  - ○ location plan (1:1000/1:1250)
  - ○ overall site plan (1:500)
  - ○ floor plans (1:200)
  - ○ elevations (1:200)
  - ○ sketches
  - ○ sections
  - ○ model views
  - ○ model analysis diagrams (massing, energy, sunlight, daylight etc.)
  - ○ key internal and external views
  - ○ indicative materials and finishes
  - ○ outline specification
- Structural strategy
- Mechanical strategy
- Electrical strategy
- Public health strategy
- Highways and civils strategy
- Landscaping strategy
- Any other discipline strategies.

### Information requirements

The information requirements for a small project should not be underestimated. The project will require the same types and formats of documents to be produced as outlined above. While the volume of information will be less on a small project, care should still be taken when planning the project to ensure that sufficient time and resources have been allowed to produce, coordinate and manage the design.

### Who manages the production of information?

Collaborative working is beginning to influence the way project teams are now organised throughout the design, procurement and delivery of a project. The evolving processes and technologies are now defining new relationships that enable the information and data produced to be shared and developed more effectively.

BS 1192 outlines a number of roles and responsibilities directly related to the management and production of CAD information throughout the course of the project. While these have been defined as a number of separate roles for the benefit of larger projects, it is likely that on smaller, less complicated schemes and in smaller offices many of these roles will be performed by just one or two individuals.

PAS 1192-2 has developed these specific roles and responsibilities to ensure that BIM processes and the information produced are properly managed.

## Project roles

**BS 1192:2007**

- Design coordination manager
- Lead designer
- Task team manager
- Interface manager
- Project information manager
- CAD coordinator
- CAD manager

**PAS 1192-2:2013**

- Information management
- Project delivery management
- Lead designer
- Task team manager
- Task information manager
- Interface manager
- Information originator

The roles highlighted are not specific to the production method and the descriptions are generally interchangeable between a BIM and CAD approach.

PAS 1192 recognises that the modelled information will be produced from different sources at different stages within the Project Programme, and while it does start to differentiate between the design team and the contractor team, again the roles remain interchangeable to ensure that a flexible approach suitable for all procurement routes can be maintained.

The information manager is responsible for developing, implementing and updating the Project Execution Plan, including the Technology and

Communication Strategies, which will set out the protocols, standards and processes. This role may be a position filled by different people over the development of the project: it could be performed by the lead designer or the project manager during the earlier stages, with the opportunity for the contractor's design manager to oversee the Technical Design and Construction stages. The lead designer will manage the delivery of the design information while also establishing the base parameters for the project, such as grid, zones and levels.

The information manager is likely to be most effective if selected from within the project team. However, the role could be maintained using an external consultant, responsible directly to the client, to provide consistency for the duration of the project.

These roles and responsibilities should be reviewed at the outset of the project and outlined within the Project Execution Plan during Stage 1, so that the appropriate disciplines can suggest additional roles that might be required at later stages.

### What format should be used to develop the Concept Design?

Developing a BIM or CAD model collaboratively during Stages 2 and 3 can be a predominantly paperless process. Model files can be digitally exchanged and developed between individual disciplines, supported with PDF drawings and documents to allow particular issues and views to be reviewed in a sketch format. These views can also be used to communicate the design to the client and stakeholders, as well as to satisfy the requirements of a stage report.

However, if a planning application is being submitted at the end of this stage or Employer's Requirements are being developed to satisfy a two-stage procurement route, then the detail required in the drawings and documents should be identified as early as possible, perhaps in the Design Responsibility Matrix, to ensure that their production and delivery is included within the scope, programme and, ultimately, the overall fee for the design services.

A PIM file will typically define the project digitally at a scale of 1:50 or above, with larger scale information being added to the model views and details in a 2D format.

With a CAD project it is not as necessary to resolve all the component interfaces to the same level of definition as a BIM model would require in order to produce a comprehensive set of GAs and details for Employer's Requirements or a planning submission. This approach may offer a more efficient way of delivering compliant information on smaller projects at this early stage.

However, CAD can be significantly less flexible than BIM when testing strategies such as fire, solar gain or heating using modelling software. 2D information needs to be completed and modelled to suit testing software and, subject to the results, redrawn and remodelled to enable the impacts of any change to be understood. Obviously, the design can be developed with the benefit of the original analysis, and this may well inform the next iteration. However, additional exercises will still be required to improve poor solutions.

Developing the Concept Design using a PIM allows the information to be tested almost simultaneously, subject to compatibility between analysis and modelling software, so that the outcomes inform the design development and reflect the key project requirements and strategies as it progresses. The benefits of early analysis at this stage will inform how to orientate a building, improve energy use, develop responsive spaces, light those spaces naturally and improve both the construction and operational life cycle costs. In addition, the speed with which strategies can be tested allows more time for the development of the design solution and so improve the quality, deliverability and information.

A PIM can be utilised to produce drawings, analyse, test coordinate clashes and schedule, cost, construct, monitor and manage the facility. However, the information contained and the levels of detail it represents can vary significantly to satisfy each of these requirements. A model produced solely for coordination and clash detection might not provide sufficient detail for the outputs required for operational use. The client and the other project team members need to understand that setting requirements and constraints on the model from the outset may restrict its use later in the process and that developing a model to perform tasks beyond its intended use may add time and cost to the project.

## Chapter summary 2

The Final Project Brief and Concept Design represent the foundation for delivering the Project Objectives and Project Outcomes. It is important to ensure that, by the end of this stage, all the briefing requirements have been addressed and, in the most part, satisfied by the preferred design, approach and brief.

The information exchanged at this point represents the basis of the final project. Any future change in the design and construction process and any significant developments in subsequent stages should be assessed against the Concept Design to determine the likely consequences on quality, cost and programme.

Stage 2 also includes the development of a number of key strategies, progressed collaboratively between the project team. The format and standards used to produce the Project Information will be in accordance with these strategies, as will the development of the Concept Design. The final information developed for exchange at the end of the stage, whether BIM or CAD drawings or models, should also be checked and verified, to ensure that it can be utilised as intended during the subsequent stages.

The Stage 2 Information Exchanges are:

| Final Project Brief
| Concept Design
| Cost Information
| Project Strategies.

Additional Information reviewed might include:

| Design Programme
| Project Execution Plan
| planning application/pre-application advice
| Sustainability Aspirations
| Risk Assessments
| Stage 2 report
| Digital Plan of Work.

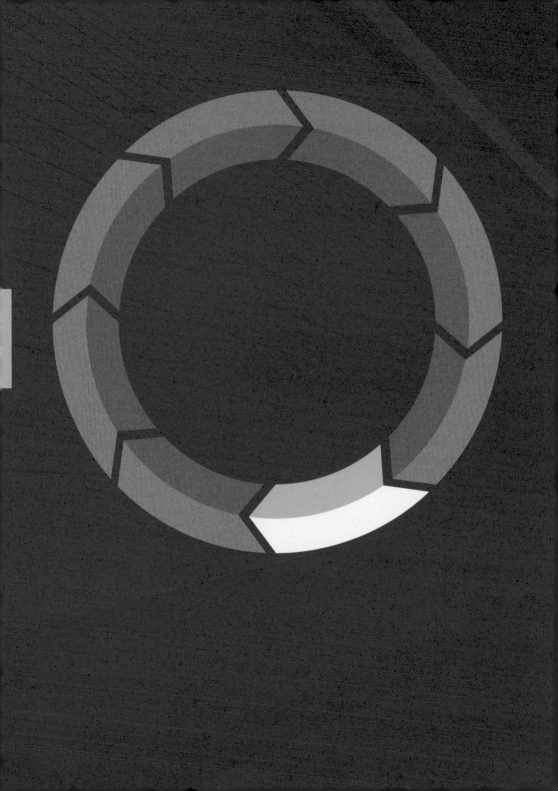

# Developed Design

# Chapter overview

During this stage, the preferred design solution will be developed to ensure that all the key building systems and interfaces have been coordinated and are aligned with the Cost Information. Subject to the procurement strategy, the responsibility for information production may shift from the design team employer to the contractor, either at the outset or at the end of this stage. In these instances, the Project Information needs to be developed to precisely describe the key areas of the design, with any remaining elements clearly identified to be completed by the contractor. Ensuring that the Project Information is structured in accordance with the defined project standards and protocols will be paramount at this stage in order to facilitate the use and exchange of data between the design team as well as any new designers that are appointed to assist in finalising coordinated design.

**The key coverage in this chapter is as follows:**

Structuring information for coordination

Composition of BIM models

Managing the coordination process

Outlining the Developed Design

What are the Information Exchanges at the end of Stage 3?

Defining key strategies and responsibilities

# Introduction

The Developed Design stage will provide the minimum level of information needed to complete a coordinated design and, if required, the submission of a full planning application. Clarity of the level of detail and information to be produced by the design team will be essential to progressing to the next stage, and the information requirements, Schedules of Services and appointments should all be suitably aligned to ensure the information is resolved collaboratively.

On many projects, the structural and MEP (mechanical, electrical and public health) systems and other design inputs included in the planning submission might not have been resolved to the same level as the architectural design. While minimising the design input in this way prior to seeking a planning approval does ensure that the client is not overexposed in relation to design fees, without the collaborative input of all members of the design team, aspects of the design may not be properly coordinated, which creates risks for the client in the Stage 4 design, and might require adjustments during Stage 4 to the Developed Design consented by the planning authority. Those proceeding on such a basis should be aware of the risks. A better approach is to develop an integrated solution collaboratively from the outset.

The Developed Design stage has been structured to support a collaborative approach, aligned across the design team, ensuring that the design and any planning application produced at this stage are fully coordinated, avoiding the need for out-of-sequence redesign exercises. Some clients will still prefer to minimise their financial risk; in these cases the Design Responsibility Matrix (DRM) can be used to highlight the minimum information requirements for the stage to minimise any associated design, programme and cost implications if a collaborative route is not progressed. Reviewing the DRM will also identify areas where design input from specialist subcontractors might be required, and, although the procurement route might not be completely resolved, their advice and technical support can be integrated within the Developed Design to ensure that the interfaces between various systems are considered and accommodated.

## What are the Core Objectives of this stage?

The Core Objectives of the RIBA Plan of Work 2013 at Stage 3 are:

| Tasks ▼ | **3**<br>**Developed Design** |
|---|---|
| Core Objectives | Prepare **Developed Design**, including coordinated and updated proposals for structural design, building services systems, outline specifications, **Cost Information** and **Project Strategies** in accordance with **Design Programme**. |

During this stage the design team will utilise the Final Project Brief and Concept Design to progress a coordinated Developed Design. All the key Project Strategies and approaches will be developed by all the appropriate design team members to align the design with the level of definition requirements highlighted within the Design Responsibility Matrix, the Project Execution Plan and the Technology Strategy.

## Structuring information for coordination

The importance of structuring and producing information in a way that facilitates collaborative working has been outlined in the previous stages. The need for a collaborative working process as set out within BS 1192:2007 and PAS 1192-2:2013 and a common data environment (CDE) has also been identified. A CDE provides a more efficient and collaborative way of working, ensuring that many of the procedural and file management tasks are integrated within the overall process, especially when the process is operated collaboratively online.

### Using a common data environment

Smaller or less complicated projects might not be able to meet the expense of a web-based management system. However, if a CDE is viewed primarily as a process and the basic principles are followed, the same efficiencies can be achieved on these projects.

Exchanging work-in-progress information and communicating changes in an email or transmittal allows the design to progress and avoids abortive work; for example, a designer who is working in isolation will be prevented from taking an incorrect solution too far. Developing and exchanging information collaboratively, in line with the four main CDE processes outlined within BS 1192, will benefit coordination and help prevent duplication, waste, errors and inconsistencies when information is shared and validated.

### What processes define a CDE?

A CDE can be utilised with a CAD or a BIM approach, to manage how data is created and exchanged by the project team members at each particular stage of development.

A CDE comprises four distinct areas (figure 3.1):

I  Work in Progress (WIP)
I  Shared
I  Published
I  Archive.

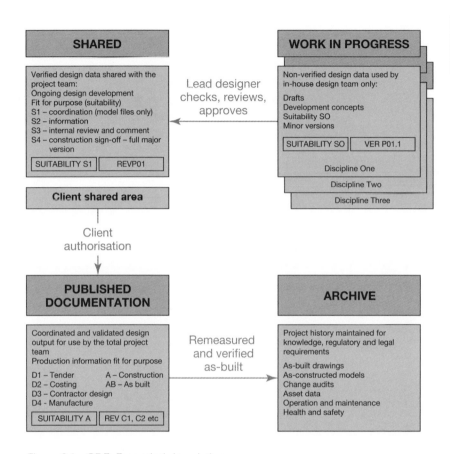

**SHARED**

Verified design data shared with the project team:
Ongoing design development
Fit for purpose (suitability)
S1 – coordination (model files only)
S2 – information
S3 – internal review and comment
S4 – construction sign-off – full major version

| SUITABILITY S1 | REVP01 |

Lead designer checks, reviews, approves

**WORK IN PROGRESS**

Non-verified design data used by in-house design team only:

Drafts
Development concepts
Suitability SO
Minor versions

| SUITABILITY SO | VER P01.1 |

Discipline One
Discipline Two
Discipline Three

**Client shared area**

Client authorisation

**PUBLISHED DOCUMENTATION**

Coordinated and validated design output for use by the total project team
Production information fit for purpose

D1 – Tender          A – Construction
D2 – Costing         AB – As built
D3 – Contractor design
D4 - Manufacture

| SUITABILITY A | REV C1, C2 etc |

Remeasured and verified as-built

**ARCHIVE**

Project history maintained for knowledge, regulatory and legal requirements

As-built drawings
As-constructed models
Change audits
Asset data
Operation and maintenance
Health and safety

*Figure 3.1    CDE: Expanded description*

Adopting these management standards is important, but doing so should not require significant changes to the processes and composition of any existing information management systems, only to the way the information is structured, validated and, most importantly, exchanged.

'Work in Progress', for example, defines the status of the information. WIP information does not necessarily need to be separated from other Project Information, it just needs to be outlined within a protocol to ensure that all those working within the project environment understand where WIP information will be stored and how it will be managed and exchanged.

Existing company or project standards, if defined, should be able to accommodate the process, and adopting these standards and methods of working should not be seen as a barrier to collaboration.

Operating a CDE will not define the format for exchanging information. This should be resolved at a project level within the Project Execution Plan and Technology Strategy. All project team members should contribute to this process so that the most appropriate approach to sharing all types of information can be determined.

Sharing information underpins a collaborative workflow, allowing the design to be developed with the most up-to-date information and, whether utilising a CAD or a BIM approach, it is important to ensure the information exchanged is appropriate and ready to be used by other disciplines. In order for information to move from one area to another within a CDE, it must pass through a sign-off procedure, which validates the content and appropriateness of the data. These gateways should be developed to suit the project specifics; however, the BS 1192 CDE process suggests that:

| lead designer reviews govern information moving from the WIP area to the Shared area
| client authorisation allows information to move from the Shared area to the Published area
| as-built information is checked and verified in order to move from the Published area to the Archive area.

The standard outlines that in order for design information to be shared it must achieve a status that is 'fit for coordination'. Data can then be uploaded to a web-based system or circulated to all members of the team via email or an alternative data exchange method in accordance with the Communication Strategy outlined within the Project Execution Plan.

To ensure that the purpose of the information exchanged is understood and that it is used as intended, the CDE introduces the idea of 'status codes' to be included within the file naming protocols (figure 3.2). These codes allow the project team members to utilise the information on the understanding that its use and impact on their own information will be restricted by its status.

As coordinating the design is an iterative process it is important to ensure that when files are updated and exchanged, especially WIP

| Code | Suitability | Models | Drawings and Documents |
|------|-------------|--------|------------------------|
| **WORK IN PROGRESS** | | | |
| SO | Initial non-contractual code. Master document index of file names should be uploaded into the extranet. | Y | Y |
| **SHARED Pre-construction sign-off. Non contractual.** | | | |
| S1 | Fit for co-ordination. The file is available to be "shared" and used by other disciplines as a background for information. | Y | N |
| S2 | Fit for information | N | Y |
| S3 | Fit for internal review and comment | Y | Y |
| S4 | Fit for construction approval | N | Y |
| **DOCUMENTATION Pre construction sign-off codes with temporary ownership by the contractor for a specified purpose. Non contractual.** | | | |
| D1 | Fit for costing | Y | Y |
| D2 | Fit for tender | N | Y |
| D3 | Fit for contractor design | Y | Y |
| D4 | Fit for manufacture/procurement | N | Y |
| **DOCUMENTATION These are sign-off codes used to state the completeness of the document for contractual purposes.** | | | |
| A | Fit for construction | N | Y |
| B | Fit for construction but with comments[A] | N | Y |
| C | Comprehensive revisions needed | N | Y |
| **ARCHIVE** | | | |
| AB | As built | Y | Y |

*NOTE 1 Codes A, B, C are referenced JCT 2005 – Major Project Sub-Contract (MPSub/G) [1] describing the sign-off of design documents for transfer to the contractor or subcontractor. This sign-off process is the same as that for manually produced drawings and is used again for CAD or electronic data/drawings. Refer to BS 7000-4.*

*NOTE 2 There is provision to extend this with project specific codes.*

*[A] For construction with minor comments from the client.*

*All minor comments should be indicated by the insertion of a statement of "in abeyance" until the comment is resolved or minor changes incorporated and resubmitted to achieve full sign-off.*

*Figure 3.2   BS 1192:2007 Table 5: Standard codes for suitability models and documents*

files, all the developments are documented. This can be within the files themselves, on the issue sheets or within accompanying emails or web-based communications or through a project schedule. This will ensure the design can be progressed collaboratively, and that abortive work associated with the developing design is avoided.

When the design achieves a particular level of definition (LOD) then the relevant model files can be composed to create drawings, including plans, sections, elevations and 3D model excerpts. With a BIM model, the drawing will represent a particular fixed view or cut through of the model itself. These model views are set out within a drawing sheet template, which will record the details of the project and the name of the drawing, its number, revision and status.

## Models

While models developed using BIM are compositions of 3D objects and despite CAD being predominantly a 2D approach, BS 1192 specifically refers to any drawn piece of information as a 'model' file. Models contain the current design and can be edited natively or when exchanged through an interoperable format.

BS 1192 defines model files as a 'collection of containers organized to represent the physical parts of objects, for example a building or a mechanical device'. It also notes that 'Models can be two-dimensional (2D) or three-dimensional (3D), and can include graphical as well as non-graphical content. This standard is based on generating, sharing, etc., model files, not just drawings. Drawings are produced when the model is complete, correct and consistent', and that 'Models can include information by reference'.

## Composition of CAD files

Figure 3.3 is based on the AEC (UK) CAD Standard for Drawing Management v.2.4 (2005) Best Practice Model of CAD File Composition and shows how model files should be used to configure design information.

## Composition of CAD files (*continued*)

Model files will contain the relevant project information, such as the grid, building layout, structural or servicing layouts, and can either be collated within a separate container file, prior to being referenced into the model space of the final drawing, or be placed directly into the drawing and arranged, cropped and reformatted to suit the drawing output.

The drawing border is referenced directly into the final drawing sheet space.

The drawing file will typically be exchanged in a non-editable format, such as a PDF file or paper hard copy.

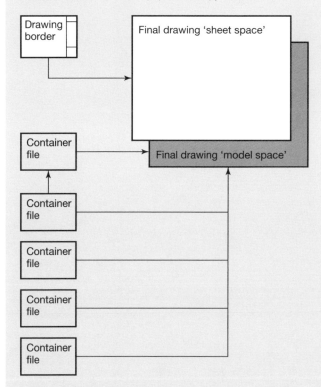

*Figure 3.3    CAD drawing file composition*

When information is no longer relevant to the project it should be archived or superseded to prevent it being used unintentionally. If the CDE process is managed through a web-based system it will tend to do this automatically, with all re-versioned files archived during the progression of the project; however, these may also be stored locally by individual disciplines.

The lead designer should retain a copy of all versioned information from all disciplines. This collection of files may become quite large, but making regular backups, following the completion of a stage for example, will help to manage the size of the archive. This process will provide a detailed history of the Project Information, as well as retaining key decision points throughout project.

Archiving can also be used to maintain a copy of all the relevant information outlined within the Handover Strategy, such as:

I remeasured as-built/constructed and verified information, as drawings and model files
I change audits
I asset data
I health and safety file, including CDM requirements
I operation and maintenance manuals
I additional deliverables (as may be required within the brief).

PAS 1192-2 develops the BS 1192 CDE processes to suit a BIM approach. It describes the shared use of individually authored models as a single source of information for any given project, used to collect, manage and disseminate all the relevant approved project documents throughout the project team. Figure 3.4 highlights how the BS CAD standards have been adapted. The basic principles of the CDE are retained, with the four main areas of information. However, the design stage CDE includes a Client Shared area, which sets out the Employer's Requirements and may include access to information on an existing building or a building to be refurbished.

It is important that the project team members understand the impacts that any design input from specialist subcontractors might have, as these areas may only be developed generically until the actual detailed information is provided. The Design Responsibility Matrix (DRM) makes provision for this input to be considered during Stage 4: Technical Design, and it can

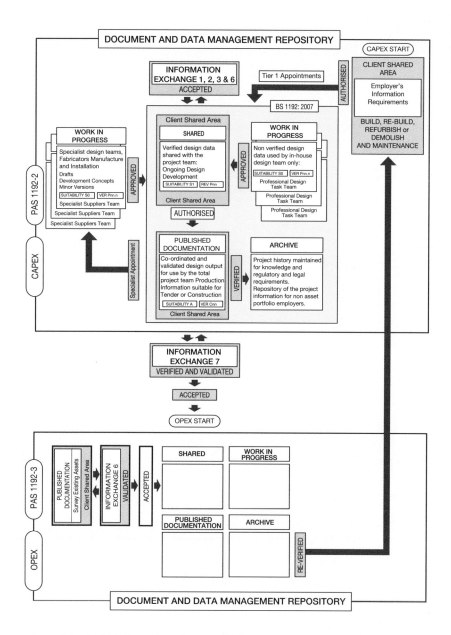

*Figure 3.4    PAS 1192-2: Extending the CDE*

| Aspect of design | | Design team | | | Contractor | | | Building contractor | |
| --- | --- | --- | --- | --- | --- | --- | --- | --- | --- |
| Classification | Title | Design responsibility | Level of design | Information exchange | Design responsibility | Level of design | Information exchange | Contractor's designed portion | Collateral Warranty required? |
| 15-05 | Substructure | | | | | | | | |
| 15-05-65 | Piling | | | | | | | | |
| 15-65-75 | Insitu concrete frame | | | | | | | | |
| 15-65-75 | Post tensioned concrete frame | | | | | | | | |
| 15-65-75 | Precast concrete frame | | | | | | | | |
| 15-65-75 | Steel frame including secondary steel | | | | | | | | |
| 20-10-20 | Suspended ceilings | | | | | | | | |
| 20-15-05 | Hard landscaping | | | | | | | | |
| 20-25-75 | Roof lights | | | | | | | | |
| 20-50-30 | Flat roof systems | | | | | | | | |
| 20-50-50 | Metal sheet roof systems | | | | | | | | |
| 20-56 | Carpets and other floor finishes | | | | | | | | |
| 20-55-15 | Screeds | | | | | | | | |
| 20-55-35 | Internal floor tiling | | | | | | | | |
| 20-55-70 | Raised access floors | | | | | | | | |
| 20-55-95 | Timber flooring | | | | | | | | |
| 25-05-60 | Panel cubicle systems | | | | | | | | |
| 25-05-65 | Relocatable partition systems | | | | | | | | |
| 25-10-55 | External masonry walls | | | | | | | | |
| 25-10-55 | Internal blockwork | | | | | | | | |
| 25-15-25 | Internal masonry walls | | | | | | | | |
| 25-15-35 | Internal stud partitions | | | | | | | | |
| 25-20 | Light steel wall framing systems | | | | | | | | |
| 25-25-10 | Fencing | | | | | | | | |
| 25-50-20 | Balustrades and handrails | | | | | | | | |
| 25-50-20 | External doors | | | | | | | | |
| 25-50-20 | Internal doors and doorsets | | | | | | | | |

4 - Technical Design

*Figure 3.5   DRM: Planning for specialist design input*

also be noted under a separate column at the end of the worksheet (as downloaded from the RIBA Plan of Work Toolbox: www.ribaplanofwork. com) (figure 3.5).

Specific areas that might require design input from specialit subcontractors should be outlined in the DRM when known. This should be possible at the outset (Stage 1) based on sector norms. This approach may affect the scope and programme of the design, but it will avoid any early abortive work or reworking of detailed information during the later stages.

Specialist subcontractor design information can be integrated directly into the model during the design, fabrication and installation process. When a single federated model is used to accommodate this, PAS 1192-2 recommends that a 'change of ownership' procedure is operated at the relevant stage to ensure the responsibility for any objects, components and assemblies that replace the design intent is always defined.

While these changes in model ownership need to be fully understood by the project team members, the design and coordination of each key interface between packages remains the most important exercise at this stage to avoid impacting the constructed quality. The project team will need to ensure that any interfaces are resolved collaboratively to reflect the intended design, performance and construction requirements.

The status codes within the PAS (figure 3.6) have been amended to include the issue of the Project Information Model (PIM), the Asset Information Model (AIM) (see Stage 6: Handover and Close Out, page 193), manufacturing models and the documentation sign-off processes.

Every project will have its own requirements and the CDE process set out within BS 1192 and PAS 1192-2 should be used as a best practice guide to outline the principles of managing and coordinating Developed Design information. The use of web-based systems may offer slightly different approaches, subject to the way the information is accessed and stored, the actual system being used and the amount of information being exchanged, but the process remains essentially the same.

| Status | Description |
|---|---|
| **Work in progress (WIP)** | |
| SO | Initial status or WIP<br>Master document index of file identifiers uploaded into the extranet. |
| **Shared** | |
| S1 | Issued for co-ordination<br>The file is available to be "shared" and used by other disciplines as a background for their information. |
| S2 | Issued for information |
| S3 | Issued for internal review and comment |
| S4 | Issued for construction approval |
| S5 | Issued for manufacture |
| S6 | Issued for PIM authorization (Information Exchanges 1-3) |
| S7 | Issued for AIM authorization (Information Exchange 6) |
| D1 | Issued for costing |
| D2 | Issued for tender |
| D3 | Issued for contractor design |
| D4 | Issued for manufacture/procurement |
| AM | As Maintained |
| **Published documentation** | |
| A | Issued for construction |
| B | Partially signed-off:<br>For construction with minor comments from the client. All minor comments should be indicated by the insertion of a cloud and a statement of "in abeyance" until the comment is resolved, then resubmitted for full authorization. |
| AB | As-built handover documentation, PDF, native models, COBie, etc. |

*NOTE 1 Additional codes S6 and S7 are highlighted.*

*NOTE 2 Status codes are provided by information originators to define how information may be used during different phases of the CDE. The SHARED suitability codes are stated as "Issued for..." but this does not infer any contractual or insurable purpose. Their purpose is to limit the reuse of the information at that stage. See also BS 1192 and Building Information Modelling – A Standard Framework and Guide to BS 1192, Richards, 2010.*

*NOTE 3 Status codes are used in connection with the gateways in the CDE. They are not related to version numbering, the levels of detail or the stages in the plan of work.*

*Figure 3.6   PAS 1192-2:2013 Table 3: Status codes in the CDE*

## Composition of Project Information Models

A BIM workflow should be structured in a similar way to the composition of CAD information. However, drawings can be extracted from each individual discipline's model or from the federated model (figure 3.7). Views from the model – plans, sections, elevations, detail callouts and 3D views – can be 'extracted' to the drawing sheet model space or to a separate drawing file, where they can be embellished with additional

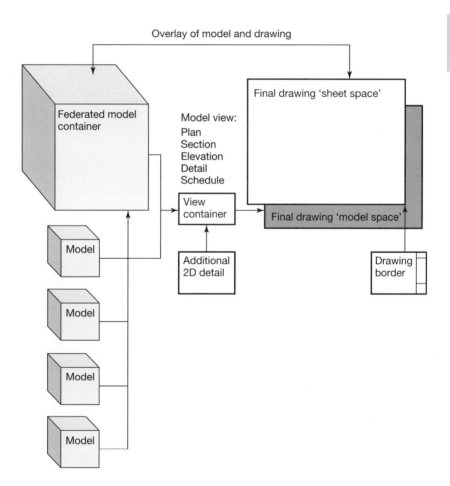

*Figure 3.7   PIM drawing file composition*

121

2D detail and setting-out and graphical information, such as:

I  dimensions
I  levels
I  setting-out points
I  additional detailed information
I  annotations
I  specification references
I  material patterning, eg insulation, concrete or brick patterns
I  line weights and line styles.

Using a simple view from the BIM model to produce the drawing is a small but very significant difference from developing information in CAD. Using information developed from views allows the drawing to retain a dynamic link with the model, ensuring that the primary information in the drawing will be updated automatically when changes are made within the model. Model updates can then be reviewed quickly in conjunction with the drawing and any embellished detail adjusted accordingly. For example, revising the setting out of an internal wall will ensure that all of its associated interfaces at, say, the floor and ceiling will update automatically within the model and any embellished information within the drawing depicting the materiality, section location, specification or additional construction detail, such as fixings, adhesives or tolerance allowances, can then be quickly amended according to the revised location.

Most information models will be drawn to reflect a level of detail that would normally be represented at a scale of 1:50. Adding more detail can create large and unmanageable model files, which are difficult for the project team members to exchange and use.

Drawings can also be produced without retaining a direct link. However, separating the drawings from the developing model can lead to coordination issues. Changes made to embellished files or the main drawings can affect the overall setting out of the components and assemblies. Therefore, if the updated information is not manually transferred back into the model, significant discrepancies may develop as the design progresses.

The principles of managing the design and of federating models are set out in figure 3.8. Each separate discipline's model is progressed locally; when it is ready for exchange it can be shared with the project team and

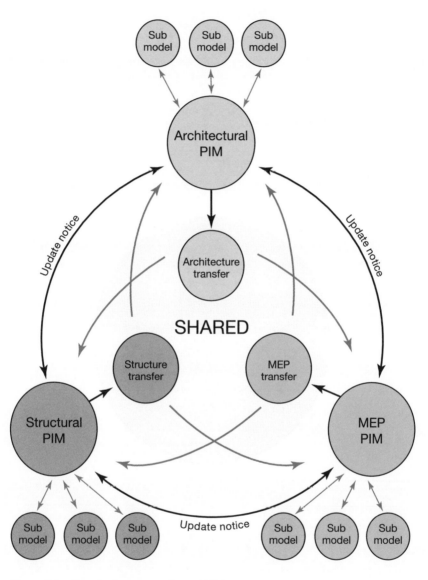

Figure 3.8    Developing a coordinated PIM

federated by the lead designer or BIM coordinator. This federated model may also exist within the Shared area of the CDE to allow all members of the project team to review their designs in context. Whenever the model is updated, the amending discipline should notify all other members of the team that revised information is available.

The model coordination process does assume that there is a centralised area for sharing models, to which all design team members have access. This Shared area is likely to be within a web-based management system, although it could simply be a synchronised location within each discipline's project folder using updated files transmitted by email, or a web-based file exchange revised manually on receipt.

Whether the project uses a CAD or a BIM approach, a CDE does require a more rigorous approach if operated manually, especially when each model file is named to include the status, version and revision. Currently, software does not identify versioning and this small change within the file name means that the main file or model will not recognise the revised information and so will not update. This can lead to the wrong file or version being used to develop and coordinate the design, turning both file management and coordination into very time-consuming manual exercises.

To overcome this issue, the Communication Strategy should stipulate that each revision or status will only be recorded on the issue sheet or within the drawing file and that the file name will remain as originally issued in accordance with the naming protocols.

A copy of the superseded information should be made before copying the new file across; the models will then overwrite and automatically be updated in all the necessary locations. The copy of the previous version will be retained in the Archive area. This will ensure that all superseded versions are archived by date and in a searchable format and, most importantly, that all information is up to date and valid.

## Managing the coordination process

Poor coordination will inevitably lead to avoidable unsatisfactory outcomes on site. Generally, the process of coordinating and resolving issues will be the responsibility of the individual disciplines, but facilitated by the lead designer. This process can be both time consuming and resource intensive. However, if the design information has been structured and developed collaboratively, then the process of ensuring that the principal components and elements are set out within the agreed design zones should become considerably easier. For example, the lead designer might suggest that the plan of a plant room is revised to provide safe and suitable access and egress from all areas while maintaining the operational and maintenance requirements of the equipment. This is a relatively simple exercise to complete at this stage, but one that might result in considerable cost and delay on site if equipment needs to be relocated once installed.

### What is clash resolution and how can it help coordination?

Using a CAD approach, coordination reviews will tend to rely on overlaying the agreed model files from each specific discipline and checking visually, 'on screen' and with hard copy drawings, to see where elements may clash.

This process should be structured and prioritised to assist the collaborative design process, with, for example, the structure being reviewed against the envelope and then internally within the floor and ceiling zones. The servicing strategies can then be overlaid, with further checks against the internal layout, structure, risers and finishes performed in these areas. This allows areas of certainty to be developed, setting out rules and zones for other disciplines.

While these checks will occur throughout the development of the design, it is likely that during the second, third and subsequent issues of the model files they will not be as rigorous as at the initial review. The designers will tend to focus only on the specific areas of change or where instances of clashes between systems or objects have occurred previously within the design. All new drawings and model files need to be completely re-evaluated, not only to ensure that items have been rectified, but also to check whether any other issues have arisen as a result or have previously been missed.

If a design develops with complex coordination constraints it can be successfully coordinated in a CAD environment, provided the information is structured to support the process and the appropriate time is set aside in the programme.

Using a CDE will ensure information is checked and validated before being issued. However, as this is also primarily a visual review by individual disciplines, the information should be structured so that disciplines can manipulate the model files to perform additional checks and validations in the context of their own designs.

Layering data allows issues to be digitally isolated (switched on and off) and reviewed in context without being obscured by other data. Separating information in this way allows the lead designer and the project team to focus on specific issues in a workshop environment, manipulating the information quickly in 'real time' to resolve the design.

Automated clash detection processes can be employed when using a BIM approach. While they are not guaranteed to detect all clashes completely, they can outline problematic issues much more quickly than if a CAD method is used, facilitating an integrated approach to the resolution of issues.

Clash resolution can be performed within each discipline's model or within the federated model, using the discipline's native design modelling software's capabilities to detect where elements and components share the same space. Alternatively, clash technology can be accessed through separate BIM integration software. Both are valid processes; however, on the basis that a collaborative team will most likely be using a range of different software and systems then the latter tends to offer a more interoperable solution. Such systems have been developed to import non-proprietary models and perform clash analysis, scheduling and detection using very specific, purpose-built tools.

The project integration software should be outlined within the Communication Strategy, identifying the format that each model should be exchanged in to accommodate the process. The Technology Strategy will identify how the information in each model is to be structured and exported, specifically for coordination purposes.

The level of development of each model is fundamental to both the checking process and the accuracy of clash detection. Generic objects, for example, might not include the information required to understand the implications of soft and sequence clashes (see page 127). PAS 1192-2 does highlight that, in order to achieve compliance at Level 2, all soft parameters should

### Automated clash detection

Typically, BIM automated clash detection processes define clashes within a number of different categories:

- Hard clashes: objects occupying the same space within the model, eg structural beams clashing with ductwork within the floor zone.
- Soft clashes: objects that have been defined with a particular tolerance or clearance requirement that has been compromised by another element, eg the designed size of the plant room failing to accommodate a maintenance zone around an air-handling unit.
- Duplicated clashes: where the same element has been created in exactly the same space more than once, eg a window object being accidently inserted twice when developing the model.
- Sequence clashes: where workflow, installation, fabrication or delivery clashes with other critical timeline issues, eg objects with programme data and attributes highlighting that floor components are to be installed before the frame that supports them.

be modelled and checked appropriately, implying that the information included should be developed sufficiently to allow this to occur.

The clash process should primarily be used to support and develop the design and not necessarily as an audit of the information produced. It can be discouraging if an initial review of a federated model identifies thousands of clashes. Managing the process and understanding the design priorities will enable the team to focus on the areas that require a collaborative input, allowing the other, less significant issues to be remedied during the development of the information.

Invariably, running an automated process on a federated model will reveal a significant number of areas to resolve, where any number of each category of clash have occurred. However, not all of them will be relevant to the coordination of the design. Using integration software will allow the clash review process to be managed using a set of basic rules outlining the specific checks to be performed, such as only reviewing the structural elements against the ductwork. It also allows other rules to be applied stating that specific instances can be ignored, such as floor boxes in an MEP model file sitting within the architectural model's floor tiles, which have not been modelled with holes to accept them.

Rules can be used to generate clash reports and schedules to:

I identify the clash owner and the discipline responsible for clash resolution
I provide an image of each clash (this may include comments)
I produce a unique ID number and name of each element
I locate the clash – by level, grid coordinates or area.

Using automated rules should not replace a visual review of each recorded clash. The process will ensure that a significant number of instances are resolved very quickly, allowing the team to focus on the key coordination issues.

Subsequent checks using integration software will be automatically managed and any new clashes will be identified accordingly. Another visual check can be performed on these new instances only, leaving outstanding issues to be managed using a clash report, which will track and report their resolution.

## Outlining the Developed Design

The Developed Design information will provide information describing all elements and components of the project and its intended construction. The information might not be in sufficient detail to construct the design, but it will be enough to allow a contractor to price and plan the works during the tender period for a single-stage design and build project, which typically uses the Stage 3 Information Exchange as part of the Employer's Requirements documentation.

The client should be advised of any risks arising from unresolved areas of the design that may impact the quality, costs or programme when finalised within the Contractor's Proposals.

### What information describes the Developed Design?

At this stage, the design intent will primarily be described by the models and drawings, but ideally the full design will be documented in a stage report. This report will, in many cases, be used as a key point of reference, demonstrating that the completed building will deliver the required Project Outcomes. It will include all the design team's information developed to satisfy the brief's aspirations as well as any statutory requirements, including Building Regulations, health and safety issues and other appropriate standards. If, during later stages, any changes are required as a result of non-compliance with standards or construction requirements, reference will be made to the Developed Design report as a record of whether the issues were considered appropriately during this stage.

The Stage 3 report should be reviewed and signed off before proceeding with the Technical Design.

## Stage 3 report – indicative information checklist

### Introduction

**General project information**

- Project team
- Summary of Final Project Brief
- Summary of Project Execution Plan
- Summary of key decisions at Stage 2

### Site Information

- Summary of key site constraints
- Summary of survey information
- Planning strategy
- Town planning analysis and implications
- Summary of discussions with statutory authorities

### Architectural

- Introduction
- History and conservation summary
- Design approach summary
- Schedule of accommodation
- Accommodation adjacency requirements
- Developed Design information (design and access statement)
- Outline specification/full specification and finishes
- Access and egress strategy (might be a separate section, depending on type, size etc.)
- Phasing/demolition
- Construction Strategy
- Planning/statutory authorities

## Stage 3 report – indicative information checklist
**(continued)**

- CDM regulations and responsibilities
- Access strategy for maintenance and operation

### Structural, civil and geotechnical engineering

- Introduction
- Summary of approach
- Design criteria/aims and objectives
- Coordinated Developed Design information
- Outline specification
- Underground drainage
- Sustainability
- Cleaning/maintenance
- Environmental
- Demolition
- Civils summary

### Mechanical, electrical and public health engineering

- Introduction
- Summary of approach
- Design criteria/aims and objectives
- Coordinated Developed Design information
- Developed outline specification
- Maintenance strategy

### Environmental strategy

- Summary of approach
- Developed Design information
- Developed specification
- Outcome of design analysis
- Transport assessment

### Specialist design strategies

- Reports produced by any specialist consultants recommended by the lead designer, including:
  - ○ acoustic or fire engineering
  - ○ arboriculture and ecology
  - ○ rights to light
  - ○ human factors
  - ○ vertical transportation

### Brief tracker and information required schedule

- Change summary

## Stage 3 report – indicative information checklist
### (*continued*)

– Summary of information required to develop the design within Stage 4
– Summary of unresolved information that may impact Stage 4 design

### Cost Information

– Stage 3 cost plan
– Commentary
– Summary of changes
– Value engineering options

### Programme and planning

– Phasing
– Design Programme
– Overall Project Programme
– Programme improvement proposals

### Construction Strategy/procurement

– Buildability
– Access and logistics
– Proposed procurement strategy
– *OJEU* notices (if applicable)

### Risk

– Summary
– Risk Register (including preliminary architectural, structural, M&E and cost risks)
– Highlight 'special' project risks

### Health and Safety Strategy

– Summary
– Integration of health and safety issues and risk into the design
– Priority health and safety issues
– Highlight 'significant' or unusual buildability and health and safety issues

### Operational and Maintenance Strategy

– Summary
– Areas that may influence the Technical Design

### Key performance indicators

– Assessment of current ability to meet KPI targets
– Action required to meet targets

## Stage 3 report – indicative information checklist
(*continued*)

Way forward

– Planning approval and discharge of any conditions
– Outline of elements not covered in Developed Design
– Outline of areas of contractor's design (summarised from the Design Responsibility Matrix)

Appendices

– Summary of project reviews
– Architectural drawings and documents, typically comprising (depending on type and size of the project):
  ○ masterplans
  ○ parameter plans
  ○ areas
– Developed Design drawings, typically highlighting the principal dimensions and gridlines, including overall internal and external spaces, floor-to-floor and floor-to-ceiling heights based on an agreed survey datum and finished grades. They may include:
  ○ location plan (1:1000/1:1250)
  ○ overall site plan (1:500)
  ○ landscaping plans
  ○ demolition plans
  ○ floor plans (1:100/1:50)
  ○ roof plans
  ○ core plan
  ○ wall types plans
  ○ reflected ceiling plans (1:100/1:50)
  ○ floor finishes (1:100/1:50)
  ○ elevations (1:100)
  ○ sketches
  ○ typical sections
  ○ detailed sections of key interfaces
  ○ planning application drawings
  ○ materials, finishes and colour boards
  ○ indicative sketches sufficient to illustrate the overall concept
  ○ model files, including study models for massing, energy, sunlight and daylight
  ○ other defined marketing material
  ○ room data sheets
  ○ door schedule – internal and external
  ○ window schedule
  ○ area schedule
  ○ finishes schedule
  ○ outline specification

## Stage 3 report – indicative information checklist
(***continued***)

- Structural strategy
- Mechanical strategy
- Electrical strategy
- Public health strategy
- Highways and civils strategy
- Landscaping strategy
- Any other discipline strategies.

This is an indicative list and should be used as a guide rather than a template.

### Managing interfaces

As highlighted in Stage 2, the amount of information required for a small project does not differ significantly from that required for other projects, other than perhaps requiring fewer documents to describe it. The coordination exercise will be just as intensive as for any other project, and may even require greater input and management by the lead designer to ensure that all the key component interfaces have been considered.

### What is required to complete the planning strategy?

If a planning application was not made at the end of Stage 2, the Concept Design will be developed during this stage to satisfy any planning requirements and pre-application advice obtained. The coordinated design will provide a definitive set of information that satisfies the consultation requirements, with the materiality and specification of the design being outlined, as well as the functional performance of the spaces. The full application may be completed during Stage 3, with further coordination developed following the submission, or it may be the final task of the stage.

All disciplines will be required to contribute to a design and access statement. The information developed for the stage report should be used to compile this and should be produced in a format that can be easily restructured to satisfy any specific requirements the statement might cover. Alternatively, the design and access statement can form the basis of the stage report if completed before the end of the stage.

The statement will include an appraisal of the design performance, its energy strategy and any environmental measures designed to satisfy the Sustainability Aspirations of the project. The structural and servicing zones will be developed in sufficient detail to allow the spatial constraints to be finalised, ensuring that any developments within the Technical Design stage will not adversely impact the form and massing, which would require further permissions or amendments.

The subsequent planning drawings will form part of the Stage 3 report issued prior to completion.

## What are the Information Exchanges at the end of Stage 3?

The Information Exchanges required for Stage 3 comprise:

I  the Developed Design
I  updated Cost Information.

For projects completed in compliance with the UK government's Digital Plan of Work (dPOW), this stage also requires a key Information Exchange:

I  UK Government Information Exchange: Data Drop 3.

### What defines the UK Government Information Exchange: Data Drop 3?

The dPOW process assumes that the design developed during Stage 2 will typically be developed by a contractor as part of a tender process prior to the Stage 2 Data Drop. The Stage 3 design will then be coordinated and developed further by the preferred contractor's design team to provide information that supports an 'agreed maximum price' at Data Drop 3. The information will include drawings, specifications and schedules. The contractor's input will also be included within the model to assist the development of manufacturing, production and assembly information during Stage 4.

Plain language questions are set out to ensure the developed design and specifications are consistent with the brief's aspirations in terms of function, cost and carbon performance. They identify a number of areas to be reviewed during this stage, including:

| the management of the PIM
| the use of protocols and standards
| how the design is coordinated
| completing coordination with existing systems
| understanding site context and information
| functional planning and adjacencies
| the required LOD
| the Construction Strategy and buildability
| the cost plan
| risk analysis
| the Health and Safety Strategy
| energy performance and assessments
| precedent studies
| statutory requirements and planning strategy.

## Defining key strategies and responsibilities

The Design Responsibility Matrix will identify what LOD each discipline will produce at this stage, as well as any areas that may require input from specialist subcontractors. In addition it will outline the inputs required to develop the Project Strategies and where additional design team members are required or retained to develop and coordinate the design. Where these are not appointed, the risks and impacts of not developing the design collaboratively should be assessed and reviewed with the client so that any risks can be identified and managed.

### What are the Key Support Tasks and Project Strategies?

The Suggested Key Support Tasks outlined at this stage comprise reviewing and updating the:

| Sustainability Strategy
| Maintenance and Operational Strategy
| Handover Strategy
| Risk Assessments
| Construction Strategy
| Health and Safety Strategy
| Project Execution Plan, including Technology and Communication Strategies
| Research and Development
| Change Control Procedures.

135

### How are design changes managed?

Changes and amendments to the Project Information following the completion of the Final Project Brief will most likely have a direct impact on quality, time and cost. Also, the later within the process that change occurs, the greater these consequences will become.

Managing change is a progressive exercise. It needs to be controlled with a robust process to ensure the implications of any new requirements are assessed and approved before being integrated into the design. The information produced within the Developed Design stage increases significantly from the previous stages and even minor changes can impact many areas of information and, subsequently, the cost and the programme.

A Change Control Procedure will be outlined within the Project Execution Plan at Stage 3, following the completion of the Concept Design. Implemented at the outset of this stage, the process will track changes to the brief and the design as they develop, identifying:

I specific details of and reasons for the change
I the impacts of the change on the design quality, cost, Project Programme and Project Strategies, such as the Health and Safety and Maintenance and Operational Strategies
I the impacts on the Project Information produced to date
I options for mitigating the impacts of the change
I who raised the change and who should review the implications prior to approval
I any risks in implementing the change
I the programme for approval.

### How are costs managed in relation to the developing design?

As the level of information and detail increases during Stage 3 the cost plan can be set out using an elemental approach to help determine whether the design is in accordance with the overall budget and identify any areas where this may be exceeded. The information should be developed to enable the cost consultant to assess:

I the estimated construction cost
I the anticipated final cost of the project

I any cost risks and their implications

I the operating costs

I any methods and areas for achieving savings.

On the basis that the Developed Design tends to define the expected cost of the project, it is important to ensure that the project team adopts a proactive approach to cost management. The Cost Information should be updated regularly with the progression of the WIP information, allowing it be used as a design tool and not as a reporting document once the design is completed. If the project is tendered at this stage, it is also important that sufficient information is provided to allow tenderers to assess the likely construction costs and to minimise the likelihood of any additional costs and claims arising as the project progresses.

## Chapter summary    3

Progressing the Developed Design collaboratively is essential to minimising errors and inconsistencies. Setting out the information using a standardised approach allows the design to progress rigorously within an iterative process, which will ensure that all aspects of the design are coordinated to the required level of definition.

Using a common data environment based on standards and protocols developed from BS 1192 should ensure that the project Technology Strategy defines the way information is structured and managed, allowing the design team to focus on coordinating the design collaboratively rather than resolving how information is formatted and exchanged.

Aligned design information allows the use of new technology to support the coordination process, particularly clash reviews and resolution, so that issues within the model files can be resolved at this stage, preventing the delays and increased cost that would occur if they were discovered on site.

The Stage 3 Information Exchanges are:

| Developed Design
| updated Cost Information.

Additional Information reviewed might include:

| planning application
| Project Programme
| Project Execution Plan
| updated Project Strategies
| Change Control Procedures
| Sustainability Aspirations
| Risk Assessments
| Stage 3 report
| Digital Plan of Work.

# Technical Design

# Chapter overview

Technical Design is structured to provide every opportunity for all of the technical and coordination aspects of the design and the integration of specialist subcontractors' design work to be concluded prior to the start on site, reducing site issues and minimising the impact on the Construction Programme and overall Project Programme. However, most methods of procurement will overlap objectives, strategies and key support tasks of Stages 4 and 5 – a process which is also under the RIBA Plan of Work 2013. It should be noted that, with the exception of Design Queries raised during Stage 5, all design work is undertaken during Stage 4.

The Core Objectives, Key Support Tasks and Project Strategies will be developed in conjunction with the design and procurement processes to ensure that no further design work, except for Design Queries arising from site – which are reviewed within Stage 5: Construction – will be undertaken following the completion of the Technical Design.

**The key coverage in this chapter is as follows:**

Providing the right information for construction

How does the specification support the Technical Design?

What information describes the Technical Design?

What are the Information Exchanges at the end of Stage 4?

Developing key Project Strategies

How is a BIM project different for subcontractors?

# Introduction

It is important that the standards and protocols used to organise the Project Information during the Concept Design and Developed Design stages are maintained throughout this stage, irrespective of who is producing the information, to ensure all the documents and files produced can be exchanged and reviewed collaboratively within the project team.

The Technical Design information, in addition to enabling the remaining coordination and integration exercises, will ultimately inform the construction of the project. It will comprise the complete design of every part and component, providing clear and concise information in support of the Construction Programme.

The timing of the tender for the construction works will vary depending on the procurement route and may take place during a number of RIBA stages. Where tendering activity takes place during Stage 4 it will need to be aligned in the Project Programme to allow sufficient time for specialist subcontractors with design responsibilities to complete their work and for that design work to be integrated into the coordinated design before the end of the stage. It is important that there is an early understanding of which aspects of the design will be undertaken by the specialist subcontractor – the Design Responsibility Matrix is the core vehicle for ensuring that this is considered in any professional services contracts. The Design Programme will also minimise the risk that changes made following the input of specialist subcontractors will impact resolved, and sometimes constructed, areas of the design.

All statutory approvals will also have been obtained by the end of this stage, including any required from the contractor prior to starting on site, such as discharging any planning conditions.

## What are the Core Objectives of this stage?

The Core Objectives of the RIBA Plan of Work 2013 at Stage 4 are:

| Tasks ▼ | 4 Technical Design |
|---|---|
| Core Objectives | Prepare **Technical Design** in accordance with **Design Responsibility Matrix** and **Project Strategies** to include all architectural, structural and building services information, specialist subcontractor design and specifications, in accordance with **Design Programme**. |

At this stage the Core Objectives develop the information sufficiently to allow the project to be constructed. The level of definition required will be defined in the Design Responsibility Matrix and the project team will need to manage when and how the final information is produced to ensure that all the construction interfaces are coordinated and integrated.

## Providing the right information for construction

The lead designer will coordinate and integrate the Technical Design. This will involve design input from the design team, but may also include specialist subcontractors developing the information collaboratively. Using the Design Responsibility Matrix to outline what information requires subcontractor design allows the design team to ensure that any interfaces with the Technical Design are identified and coordinated prior to the issue of construction information for use by the specialist subcontractors.

On large or complex projects, where the post-tender design input might be significant, the contractor may also use a design manager to assist the design team to ensure that any specialist subcontractor input is aligned with the design intent prior to release. The design team should also regularly review the coordinated Technical Design information to ensure it is aligned with the Construction Strategy and other Project Strategies.

### What information defines the Technical Design?

The design team, including any specialist subcontractors, will manage, coordinate and close out any design issues. The Technical Design information will typically comprise:

I   location drawings
I   assembly drawings
I   component drawings
I   specifications
I   schedules.

As this information will be used by the contractor to construct the design it important to ensure that the dimensions are sufficiently accurate and the specification and design are complete.

### Drawn documentation

Location drawings include the general arrangement plans, floor and ceiling plans, sections and elevations outlining the general setting out of the project. They will include the project grid and cross-references to assembly and component drawings where further detail on key areas can be found. The level of annotation on location drawings should be restricted to spatial descriptions, such as building and room names, and component references for items such as doors and windows. Grid dimensions and setting-out points should be illustrated; however, any detailed setting out should be recorded on larger scale drawings specifically designated for that purpose.

Assembly drawings show how elements are integrated and how the building is constructed. They consist of large-scale plans in whole or part and building sections, and can also include elevations and 3D projections. This information will highlight the construction components and key dimensions and can provide references to more detailed information on component drawings.

It is important to structure assembly documentation properly to ensure that the information remains clear and uncluttered. Both CAD and BIM approaches enable drawings to be set up very quickly, and the actual composition of the drawing sheet should be carefully considered to isolate a particular model view, such as a plan or section, or to ensure that only a single drawing is placed on each sheet, so that the requirements are clearly communicated. Larger projects may require the information to be structured using defined zones in plan or section. These should be consistently maintained throughout all of the drawn information to ensure it can easily be cross-referenced on site and by other members of the project team.

CAD and Project Information Model (PIM) files are created at a scale of 1:1. However, the scale to which drawings are produced is also important, as will be the sheet size. While information should not be compromised simply to fit a drawing onto a particular sheet size, these documents will be utilised on site to construct the project and so will need to be manageable. For example, using A1 sheets for drawn information will also allow drawings to be printed at A3 for reviewing at a corresponding scale. A0 sheets should only be used after careful consideration and for particular types of drawn information, as viewing the full drawing on site, in wet and windy conditions, can be difficult, and drawings of that size are often folded to focus on a particular issue, obscuring other relevant information.

Component drawings tend to describe the size, shape, material specification and assembly of the component as required by the manufacturer in order to produce it, such as a door or window. CAD processes, however, allow information to be produced more efficiently and accurately; component drawings can now be created by referencing a location or assembly drawing into a sheet at a larger scale, with any additional detail and annotation added as required.

Component information has typically set out details for fabrication and manufacture. However, producing the information in context within the model file or BIM model means that all the interfaces with the surrounding construction will be highlighted. This enables both the manufacturer and contractor to understand a component's:

I performance requirements
I construction sequence
I buildability
I finished quality.

### Producing detailed information

Smaller projects can tend to produce more prescriptive design solutions than larger projects, which utilise specialist subcontractors to complete significant areas of the design. While this allows greater control over the quality and interfaces of the components and systems designed, it may result in the production of more assembly and component information, to ensure the project requirements are outlined for tender and construction. The scope and Design Responsibility Matrix should be developed to ensure this has been allowed for so that the appropriate amount of information can be coordinated and resolved.

## How does the specification support the Technical Design?

Specifications represent a written description of the quality of the project. They are contract documents to be read in conjunction with the Technical Design. Producing a specification is very much part of the design process and the time required to complete a comprehensive, well-written document should not be underestimated.

## Specifications

Specifications are key documents defining:

- product materials
- fabrication requirements
- quality
- performance requirements
- sample requirements
- installation requirements
- tolerances
- standards
- testing requirements
- level of workmanship.

While the Project Information should be cross-referenced with the specification, its content should not be reused to annotate or describe components within the other project documents. References should only highlight the specific work section and clause, eg F10:110 or 25-10-55/150. This avoids any repetition of notes and descriptions within the Project Information, which can lead to conflicts, delays and additional costs should a change occur.

## Structure of a specification

Specifications are structured in three main sections:

- Preliminaries
  General information for the contractor.
- Work sections
  Component information, divided into specific areas of work defined using CAWS (Common Arrangement of Work Sections for Building Work) or, more recently in BIM projects, Uniclass 2 references (developed from Uniclass table J, which incorporated CAWS – see www.cpic.org.uk/uniclass/ for more details).
- Schedules
  Lists setting out specific components, such as doors, windows and finishes.

The UK government's Digital Plan of Work classifies objects using a mapping system that allows the objects to be referenced using NBS (National Building Specification) codes and the new Uniclass 2 system.

The levels of definition, detail and information used within specification documents can vary significantly, depending on whether the specification is prescriptive or performance based. Most specification documents will include a combination of descriptive and performance clauses, with the areas that are important to the design intent being specified prescriptively, using proprietary products, and other areas remaining less defined, to allow the contractor to use its expertise to add value.

### Specifications

Specifications are generally developed using proprietary specification-writing software, such as *NBS Building* or *NBS Create* (produced by NBS, based on the National Building Specification). For smaller domestic projects, *NBS Domestic Specification* can be used. This is an online specification tool for the provision of single project-specific specifications.

With the development of BIM there are now opportunities to link the specification data and requirements to objects and components within the BIM model. Direct correlations can be made between the performance data and the geometry, all referenced from a single location, preventing unnecessary errors and conflicts following design development and changes. These links will ensure that the model objects, drawn information and specifications are cross-referenced appropriately, allowing the qualitative and workmanship specifics of assemblies and components to be costed and controlled.

### Types of specification

### Prescriptive specifications

Prescriptive specifications represent a completed design, with little or no flexibility for the contractor to propose alternatives. In particular, the information ensures that both the quality and performance of components will be as designed and coordinated. For example, ABC Brand, Model 123, referenced to manufacturer's recommendations.

### Descriptive specifications

Descriptive specifications can outline the project requirements without using proprietary products or identifying specific manufacturers and suppliers, on the understanding that the contractor will propose acceptable equivalents as the Technical Design solution develops.

### Performance specifications

Performance specifications set out the requirements for each system or component and the criteria through which proposals will be assessed and validated. This allows the contractor scope to improve the buildability of the project and, in some cases, offer a better, more cost-effective solution, especially on less complicated projects. For example, an $X$ dimension by $Y$ dimension material finished in colour $A$, with references to British Standards and BBA certificates.

BBA Certificates are issued by the British Board of Agrément, an independent UK organisation that reviews construction products, systems and installers. Agrément certificates are issued for products, building materials and processes to confirm they perform as required.

Performance specifications should always maintain references to quality and workmanship, through the use of standards and testing requirements, and, in addition to the specified performance, should also reference compliance with appropriate statutory requirements.

### Technical specifications

British Standards (BS) and European Standards (BS EN) are technical specifications or practices that outline guidance for the making of a product or carrying out of a process.

### Validating digital objects

Digital objects and data are now becoming readily available and can be sourced from previous projects, online BIM libraries and manufacturers' and suppliers' websites. This information is likely to contain generic or very specific specification data, and so will need to be reviewed and verified by the design team before it can be finalised and used within the project specification.

## What information describes the Technical Design?

Any developments within the Project Information at this stage can be collated and summarised within a separate report. This report will outline any differences from the Developed Design as a result of the completion of the Technical Design. This might accommodate changes, revisions or specification amendments that have occurred as a result of specialist subcontractor input, value engineering exercises or further coordination exercises.

The Technical Design information will primarily comprise documents developed to communicate and describe the construction requirements. It will include supporting reports and strategies developed to accommodate changes through specialist subcontractor design input or development of the Contractor's Proposals.

The Technical Design information can also be utilised to complete applications for statutory approvals, such as Building Control, and the resolution of planning conditions. If the project has procured the services of an approved inspector then the information can be periodically reviewed throughout this stage to ensure compliance with the Building Regulations prior to a formal application. The timing of these submissions should be programmed to ensure the design is not progressed too far without validation, avoiding unnecessary abortive work should significant changes be required.

### Stage 4 Project Information checklist (indicative)

Typically at this stage the Project Information will comprise drawings, schedules, specifications and schedules of work, including:

Stage report
- Changes
- Value engineering

Location information
- Location plans and site plans showing the complete development, including parking, levels, landscaping
- Plans showing any required zoning and phasing
- General arrangement floor plans, sections and elevations

## Stage 4 Project Information checklist (indicative)
### (continued)

- Finishes information in plan, elevation and section
- Project Information Model

### Assembly information
- Floor plans at each level with setting out at a scale of 1:50 or larger
- Roof plan showing falls, gutters, rainwater heads and downpipes and penetrations
- Reflected ceiling plans at each level showing coordinated lighting and services fixtures with setting out at 1:50, including ceiling construction and support
- Large-scale external elevations
- Large-scale internal elevations
- Detailed sections, typically at 1:20 and 1:10
- 1:20 details of access cores, stairs and ramps
- 1:20 details of other key areas

### Component information
- 1:5, 1:2 and 1:1 construction details through all typical and atypical locations

### Schedules and reports
- Internal finishes schedule
- External finishes schedule
- Door schedule
- Window schedule
- Ironmongery schedule
- Sanitary schedule
- Furniture, fixtures and equipment (FF&E) schedule
- Statutory signage schedule
- Room numbers and project signage schedule
- Room data sheets
- Schedule of areas
- Maintenance report, including access requirements
- Fire strategy report, including separation, compartments and the escape strategy (including travel distances and staircase widths)

### Additional Project Information
- Coordinated builder's work in connection (BWIC) with mechanical, electrical and public health installations (MEP)
- Structural drawings and schedules
- Mechanical services drawings, schematics and schedules

### Stage 4 Project Information checklist (indicative)
(**continued**)

- Electrical services drawings, schematics and schedules
- Public health services drawings, schematics and schedules
- Civils drawings and schedules
- Landscaping drawings and planting schedules
- Other specialist design drawings and schedules

Specifications
- Architectural specification, including preliminaries and all trade sections
- Samples boards, samples, materials still requiring approval by client
- Schedules of work
- MEP systems specifications
- Structural specifications
- Civils specifications.

This is an indicative list that can be used as a guide rather than a template.

## What are the Information Exchanges at the end of Stage 4?

The Information Exchange required at the completion of Stage 4 comprises:

I   the Technical Design.

The UK government's Digital Plan of Work is structured to suggest that, as the early appointment of the contractor is preferred, the contractor will develop and manage the Project Information from an early stage. As such, much of the design intent would have been finalised at Data Drop 3, during the Developed Design stage. In fact, the onus on the contractor to design and construct the project is such that no further Data Drops are required by the client until after construction has been completed at the end of Stage 5.

While no UK Government Information Exchanges are defined for the Technical Design, there are a number of plain language questions that, as with previous stages, require responses to inform the finalisation of the design and the progression to the Construction stage. These explore issues such as:

I management of the PIMs
I compliance with protocols and standards
I compliance with statutory requirements
I review of existing Site Information
I review of performance and operation requirements
I review of maintenance requirements
I understanding of operational and life cycle costs
I facility management requirements
I scope of the operating and maintenance manuals
I review of the Project Information, including a specification,
I validation of the PIMs.

## Developing key Project Strategies

While the Project Strategies will be updated to ensure they are coordinated to align with the Technical Design, it is important that they support the development of the Project Information to include coordinated information from the design team and any specialist subcontractors. Any changes from the Developed Design, perhaps as a result of adding further detail, accommodating the design of others or responding to a value engineering exercise, should be recorded accordingly and reviewed against the Final Project Brief to ensure that the original intent is maintained. The Change Control Procedure should also be utilised to ensure that any developments are fully explored and agreed before incorporating the revisions.

The procurement route will have a considerable impact on when information is produced during this stage. With a design and build approach it is possible that the design team will alter or change. The structure and format of the information must be sufficiently robust to be used and developed by any new party and the standards and protocols used should be clearly outlined in the supporting documentation.

### What are the Key Support Tasks and Project Strategies?

The Suggested Key Support Tasks outlined at this stage include reviewing and updating the:

I Sustainability Strategy
I Maintenance and Operational Strategy

I Handover Strategy
I Risk Assessments
I Construction Strategy
I Health and Safety Strategy
I Project Execution Plan, including Technology and Communication
  Strategies
I Change Control Procedure.

These strategies will inform and develop the status of the Project Information as the Technical Design is developed. The information protocols for this stage will be outlined within the Technology Strategy and the Communication Strategy, and will be closely aligned with the procurement strategy and Construction Strategy. This will ensure that the collaborative approach developed in the earlier design stages is maintained, irrespective of who produces the information. Any design input from specialist subcontractors will need to adhere to these protocols and standards to ensure that the quality and coordination of the design, including any key details and interfaces between design packages, can be coordinated and completed.

### How can the Health and Safety Strategy improve the Project Information?

The Project Information will be used extensively on site. Therefore, each 'printed drawing' offers considerable potential for communicating any residual health and safety risks outlined within the Project Strategies, including advice on finding further information, if required. Any member of the project team involved in producing design information for construction, should integrate this health and safety information with their published Project Information.

A SHE (safety, health and environment) box (figure 4.1) can be used to communicate any residual risks relating to construction. The box can be added to the design information during the Developed Design or Technical Design stages, subject to reviews by the project team. The SHE box should summarise only significant outstanding issues: long generic lists should be avoided as they will tend to hide the more noteworthy issues.

The SHE box information should be reviewed following the completion of construction and retained on any 'As-constructed' Information produced

| SAFETY, HEALTH AND ENVIRONMENTAL INFORMATION |
|---|
| In addition to the hazards/risks normally associated with the types of work detailed on this drawing, note the following significant risks and information: |
| CONSTRUCTION<br><br>Ensure temporary supports remain in place until frame is complete<br><br>Encapsulated asbestos in existing ceiling void |
| MAINTENANCE/CLEANING |
| DECOMMISSIONING/DEMOLITION<br>Encapsulated asbestos in existing ceiling void |
| Risks listed here are not exhaustive, refer to the project CDM risk assessment register for full details: [*Insert document reference*] |
| For additional information relating to use, cleaning, maintenance and demolition, see the Health and Safety file |
| It is assumed that all works will be carried out by a competent contractor working, where appropriate, to an approved method statement |

Figure 4.1   Typical SHE box

within Stages 5, 6 and 7. This will ensure that any remaining risks not mitigated by the contractor, especially those that are user or future maintenance related, are highlighted.

## How is a BIM project different for subcontractors?

Many specialist subcontractors use CAD, or perhaps BIM, to produce their information. However, as much of this information is produced solely for internal purposes, such as fabrication information, these aspects do not need to conform to the BS 1192:2007 standards and protocols. This does not necessarily affect the way the information is coordinated or utilised as often the design will be validated and checked by referencing the information into the design intent model files or by reviewing hard or soft copy 'drawings'. However, for subcontractors developing a project using a BIM approach, utilising the project protocols and standards will be paramount to maintaining and developing the Project Information.

It is likely that all the components within the design will be developed within the federated model, and while the output information won't necessarily change, the design and interfaces of all the surrounding trades will need to be coordinated to the same level of definition as each different package is integrated.

Each specialist subcontractor will need to understand the surrounding model elements and coordinate with them. Equally, these adjacent designs will need to have been finalised or developed in conjunction with the other specialist subcontractors. This implies that the scopes and procurement of all related packages will need to be aligned to ensure the Project Programme is maintained and that work is not continuously revisited as other inputs are developed.

### Specialist subcontractors' input to the BIM model

Areas within the BIM model that require specialist subcontractor input, such as a curtain wall, could be created as a separate model and federated into the main design. This will allow the specialist subcontractor's model to replace the design intent as it is developed, ensuring that:

- interfaces are highlighted and managed by the correct disciplines
- ownership is maintained throughout construction
- fabrication drawings can be produced from a federated solution
- the 'as-constructed' model is updated correctly.

In most design and build projects the production of information is managed to suit the Construction Programme. Often, specialist subcontractors are not appointed until they are required to progress the design. While this may accelerate the Design Programme activity, it does not benefit the collaborative development of Project Information. Often, a number of

**4 - Technical Design**

| Aspect of design | | Design team | | | Contractor | | | Building contractor | |
| --- | --- | --- | --- | --- | --- | --- | --- | --- | --- |
| Classification | Title | Design responsibility | Level of design | Information exchange | Design responsibility | Level of design | Information exchange | Contractor's designed portion | Collateral Warranty required? |
| 15-05 | Substructure | | | | | | | | |
| 15-05-65 | Piling | | | | | | | | |
| 15-66-75 | Insitu concrete frame | | | | | | | | |
| 15-66-75 | Post tensioned concrete frame | | | | | | | | |
| 15-66-75 | Precast concrete frame | | | | | | | | |
| 15-66-75 | Steel frame including secondary steel | | | | | | | | |
| 20-10-20 | Suspended ceilings | | | | | | | | |
| 20-15-05 | Hard landscaping | | | | | | | | |
| 20-25-75 | Roof lights | | | | | | | | |
| 20-25-30 | Flat roof systems | | | | | | | | |
| 20-30-50 | Metal sheet roof systems | | | | | | | | |
| 20-55 | Carpets and other floor finishes | | | | | | | | |
| 20-55-15 | Screeds | | | | | | | | |
| 20-55-35 | Internal floor tiling | | | | | | | | |
| 20-55-70 | Raised access floors | | | | | | | | |
| 20-55-85 | Timber flooring | | | | | | | | |
| 25-05-80 | Panel cubicle systems | | | | | | | | |
| 25-05-65 | Relocatable partition systems | | | | | | | | |
| 25-10-55 | External masonry walls | | | | | | | | |
| 25-15-25 | Internal blockwork | | | | | | | | |
| 25-15-25 | Internal masonry walls | | | | | | | | |
| 25-15-35 | Internal stud partitions | | | | | | | | |
| 25-15-55 | Light steel wall framing systems | | | | | | | | |
| 25-20 | Fencing | | | | | | | | |
| 25-25-10 | Balustrades and handrails | | | | | | | | |
| 25-50-20 | External doors | | | | | | | | |
| 25-50-20 | Internal doors and doorsets | | | | | | | | |

*Figure 4.2  Design Responsibility Matrix Stage 4: Technical Design*

critical interfaces will have to be redesigned to suit specialist subcontractor proposals developed in isolation. For example, the external cladding solution might be procured independently from the curtain walling and, if developed separately, late changes may be required to ensure they are successfully integrated with the coordinated design.

BIM is a collaborative process and its benefits are derived from the project team working together to achieve an integrated solution. This approach will deliver efficiencies in programme, cost and quality within the final design, improving considerably on current practices.

Reviewing the Design Responsibility Matrix as the implications of the sequencing and interfaces become clearer can allow for some of the key specialist subcontractors to be procured early in the Design Programme, to help minimise the design risks and benefit the overall design. The input required from specialist subcontractors will be outlined in the Design Responsibility Matrix during Stage 1 and may be updated prior to the Building Contract being awarded. As outlined in Stage 3, the required Project Information can be set out in a number of ways (see figure 4.2). The matrix allows the core responsibility to be identified, ie design team member or specialist subcontractor, and notes can be added to highlight key issues or interfaces between related trades and packages.

The early appointment of specialist subcontractors can also ensure that the design is coordinated to improve maintainability and access. This will improve the management and operation of assets and systems during Stage 7. In particular, the actual performance and geometric details of each system can be defined within the PIM to allow the layout and location to be reviewed and the best approach to maintaining key elements developed accordingly.

### What protocols should be used to review and verify information?

The procurement strategy, Project Programme and Design Responsibility Matrix define the requirements for any specialist subcontractor design, which is to be set out and managed using the project standards and protocols within the Project Execution Plan. The subsequent Project Information will be reviewed and verified in accordance with the design intent as issued by the design team. All comments on the quality and content of the information should be documented clearly, collated by the lead designer and circulated to provide feedback and direction to the

**DOCUMENT COMMENTS**

Job Name:  Project Document Reference          Job No:  Unique Project Number

| To: | Originator | | Date: | Date of Issue |
|-----|-----------|--|-------|---------------|
|     | Discipline | | | |
| FAO: | Key Contact | | From | Reviewer |

We have reviewed the drawings and/or other documents listed below.
The purpose of this form is to record our comments on these documents.
It should be noted that the Architect's review of contractor's drawings is solely to establish general acceptability of appearance, and general compliance with the design intent or contract documents.  Comments, or lack of comments, on Contractor's Proposals shall in no way relieve him of his obligation to comply with the Contract Documents and with the terms of any separate Agreement with the Employer.  It is specifically noted that a complete dimensional check has not been undertaken by the Architect, and this, along with his other obligations, remains the Contractor's responsibility.

| Document(s) | Rev | Status | Comments |
|-------------|-----|--------|----------|
| Project Document Number | Revision | Status | 1. Outline the first comment (this issue should be clouded on the document and annotated with a number 1)<br>2.<br>3. |
| Project Document Number | Revision | Status | 1. Outline the first comment (this issue should be clouded on the document and annotated with a number 1)<br>2.<br>3. |
| Project Document Number | Revision | Status | 1. Outline the first comment (this issue should be clouded on the document and annotated with a number 1)<br>2.<br>3. |

Status:
A – No comments
B – Comments made, please resubmit revised drawings. Contractor may proceed at own risk.
C – Comments made, please resubmit revised drawings or additional information prior to proceeding.

Signed          Reviewer

*Figure 4.3   Typical document comment sheet*

originator and other appropriate disciplines so that they also understand the impacts.

Comments will typically be made by 'marking up' the issue directly on the drawing, schedule or specification. This can be completed manually on a printed copy or digitally within an editable PDF copy. If a common data environment (CDE), such as a web-based document management system, is being used this, subject to the system used, may also offer a 'mark-up' protocol for recording comments on the Project Information. In addition to the specific comments, the mark-up should also record the reviewer's recommendation as to the status of the information, based on those set out in BS 1192:2007 and PAS 1192.2013.

The process of marking up information can, if there are a number of details to comment on, tend to clutter the drawing sheet and confuse the message. To avoid misinterpretation, all the information applied in the mark-up should also be included on a separate document. This document will describe the full issue, including drawing number, revision, date of issue, each specific comment and the overall commented status. The comment sheet should be numbered as a project document accordingly and, most importantly, it will include the status definitions, if different from the British Standard, as used by the auditing discipline (see figure 4.3). A CDE should create a recorded copy of the information commented on, which will remain associated with the information throughout its use as a record of its development.

### Status codes

Both BS 1192:2007 Table 5, Standard codes for suitability models and documents, and PAS 1192-2:2013 Table 3, Status codes in the CDE, outline, some typical definitions for status codes (see Stage 3: Developed Design, figures 3.2 and 3.6).

The BS 1192: definitions are:

A   Fit for construction
B   Fit for construction but with comments*
C   Comprehensive revisions needed

*For construction with minor comments from the client.

## Status codes (*continued*)

The standard also notes that 'All minor comments should be indicated by the insertion of a statement of "in abeyance" until the comment is resolved or minor changes incorporated and resubmitted to achieve full sign-off'.

The PAS 1192-2: definitions are:

A   Issued for construction
B   Partially signed-off: For construction with minor comments from the client. All minor comments should be indicated by the insertion of a cloud and a statement of 'in abeyance' until the comment is resolved, then resubmitted for full authorisation.

It is important to ensure that the originator of the information and its reviewer are aware of the overall process and the definition of each status. It is also likely that the reviewed information may not have been completely checked against all other systems and interfaces: in the majority of cases the originator will be responsible for finalising the information.

The example comment sheet shown in figure 4.3 utilises a similar approach to the BS 1192 definitions, outlining the status as:

A   No comments
B   Comments made, please resubmit revised drawings. Contractor may proceed at own risk.
C   Comments made, please resubmit revised drawings or additional information prior to proceeding.

It also includes a number of caveats, which highlight the extent of the review completed.

All 'marked up' information should provide a direct reference on the document, or a cross-reference to the document comment sheet, in order to maintain a link between the comment status and its definition and, therefore, the intent implied by the reviewer. This will avoid information progressing incorrectly and without coordination.

## Chapter summary  4

The success of the Technical Design stage will be dictated by the robustness of the Design Programme and Construction Programme. Sufficient time must be allowed for the completion of the tender information, the contractor's appointment and the collaborative finalisation of the Project Information, including any specialist subcontractor input.

The responsibility for the production of the Technical Design may change within this stage, especially following the appointment of the contractor. It is therefore important that all the information modelled, drawn, scheduled or specified is structured for development and coordination by any member of the project team, irrespective of process and software.

Utilising the knowledge and skills of the contractor prior to their appointment and developing solutions with selected specialist subcontractors before the Building Contract is awarded will improve all aspects of the design. However, if the contractor is appointed before the design is finalised then it is likely that best value will be achieved through reviewing the programme and working collaboratively, to ensure that the Technical Design is completed prior to the start of Construction. This will reduce errors, save cost and time and improve the finished quality of the building.

The Stage 4 Information Exchange is:

⏐ the Technical Design.

Additional information reviewed might include:

⏐ Project Programme
⏐ Project Execution Plan
⏐ updated Project Strategies
⏐ Change Control Procedures
⏐ updated Cost Information

| submission for approval under the Building Regulations
| Sustainability Aspirations
| Risk Assessments
| Stage 4 report/contract information
| Digital Plan of Work.

# Stage 5

# Construction

# Chapter overview

Stage 5: Construction is concerned solely with the completion of the project, both on and off site. All of the design inputs from the design team and specialist subcontractors will have been completed at Stage 4 and the design team will now be focused on the resolution of Design Queries and developing the 'As-constructed' Information, which will be used to inform the project 'In use' during Stage 7.

**The key coverage in this chapter is as follows:**

What are the Information Exchanges at the end of Stage 5?

Incorporating Design Queries from site

What are the Key Support Tasks and Project Strategies?

Completing 'As-constructed' Information

# Introduction

The completed design may be the starting point for the Construction stage; however, some procurement routes and Project Programmes require elements of the design to overlap with construction. The design team may also be required to respond to any site-related Design Queries arising as a result of ambiguities, tolerances and interfaces between different aspects of the design.

The Project Programme will help identify whether the scope of the design will extend into the Construction stage. The design team members' professional service contracts and Schedules of Services will highlight any additional requirements, including monitoring the progression or quality of the works or the preparation of 'As-constructed' information.

The Project Execution Plan and a number of the key Project Strategies, such as the Construction Strategy and Handover Strategy, will identify who is responsible for producing 'As-constructed' Information and how it will be recorded on site, whether using contractor's mark-ups, progressive surveys of completed work, measured surveys of the final project or just final construction issue information from the design team caveated appropriately. The Project Execution Plan will also outline how the data will be updated and the format in which it should be developed at the end of the stage, for use by the client, the operator and the facilities management (FM) team. This data will also include the measured performance of any commissioned systems and any adjustments made to achieve the designed output. This will ensure that the data reviewed over Stage 6: Handover and Close Out is accurate, providing an appropriate starting point for the operation of the building in use.

## What are the Core Objectives of this stage?

The Core Objectives of the RIBA Plan of Work 2013 at Stage 5 are:

| | |
|---|---|
| | **5** ◯ |
| | **Construction** |
| **Tasks** ▼ | |
| Core Objectives | Offsite manufacturing and onsite **Construction** in accordance with **Construction Programme** and resolution of **Design Queries** from site as they arise. |

During Stage 5, the Core Objectives assume that the design is complete, and outline how the Project Information can be developed to accurately reflect any 'as-constructed' changes that occur during construction. The programming and procurement associated with modern construction is likely to result in the Design Programme overlapping with the Construction stage, requiring further coordination and clarification from the design team and generating additional site-specific issues and Design Queries.

Integrating any residual Technical Design issues and incorporating solutions to Design Queries will be critical to developing accurate 'As-constructed' Information. This will allow the project team to evaluate whether the design has been built in accordance with the Final Project Brief and the Building Contract. However, its real value will be in informing the operational phase of the project. The project team should ensure that the data is developed and updated as accurately as possible in order to realise the performance and efficiencies that will be required throughout the In Use stage.

# What are the Information Exchanges at the end of Stage 5?

The Information Exchange required at the end of the Construction stage comprises:

I the 'As-constructed' Information.

There is no UK Government Information Exchange at Stage 5. However, this stage does have a number of plain language questions that require appropriate responses to close out the Construction stage and progress to Stage 6: Handover and Close Out. Typically, these responses will identify information outlining the:

I safe management of the site
I potentially hazardous areas
I mitigation of risks
I management of change
I model-based information and samples
I commissioning sequence
I outstanding contractual issues
I accuracy of information.

## Incorporating Design Queries from site

Design Queries are typically the result of poor, inaccurate or incomplete information arising from a number of conditions, such as:

I manufacturers' tolerances
When fabricated components are aligned, standard tolerances from each manufacturing process may mean that finishes surrounding these may need to be adjusted to accommodate the differences, eg understanding how a window abutting a steel frame may affect the reveal width if maximum tolerances have been exceeded.

I installation tolerances
Installing components to accept other elements may require a margin to allow for manufacturers' tolerances and workmanship, such as installing a prefabricated metal door set within a skewed brick opening may result in interface and coordination issues that require further resolution.

| design discrepancies

Uncoordinated or out-of-date design information and late changes can cause interface and coordination discrepancies that require further review during the Construction stage.

| the interpretation of information

Construction based on interpretation and assumption can lead to significant coordination issues, so any missing or unclear information will need to be resolved through Design Queries in order for the works to progress appropriately.

The resolution of Design Queries may lead to changes to design information. It is important to ensure that their resolution is recorded and incorporated within the Project Information.

## How should Design Queries be reviewed?

Programme and cost pressures throughout the Construction stage can compress the time allowed to resolve site issues and Design Queries. Often, to ensure works progress, a review of the relevant information may be limited to the components immediately impacted within the construction sequence and, as a result, the wider implications of the change may be overlooked.

For example, resolving floor level discrepancies in a cast slab to accommodate the fixing of a number of column bases might appear relatively simple to overcome. However, if the bases are all re-levelled differently to accommodate the slab, the steelwork installation will progress, but the implications for the subsequent floor levels, any elements fixed to the columns, their interfaces, services installations and the floor and ceiling finishes will all need to be reviewed before progressing. The cumulative effects of a simple solution to overcome a site issue can create significant issues for other elements. Likewise changing the construction of an external wall to suit, say, sequencing and costs, would appear a simple amendment to review. However, in addition to the obvious geometric issues, such as the width of construction, other implications, including aesthetics and planning, fire strategy, environmental performance, structural requirements and acoustic requirements, will also need to be reviewed before a coordinated response can be prepared.

Minor changes to key environmental and structural systems can have considerable implications. It is better for the contractor to fully track, test and record the consequences of a change when it first arises, as opposed to waiting until the completion of the installation to review it, especially if rectification work is required as a result.

Reviewing Design Queries takes time. It is important to ensure that the project team allows adequate resources to consider each query, and also the appropriate time to update the Project Information regularly so that the wider implications can be assessed.

All site-developed solutions, sketches, notes, drawings, tests and surveys should be retained and, where necessary, incorporated within the Project Information to enable the validation and verification of the final 'As-constructed' Information.

### What information does the client need to move forward?

'As-constructed' Information carries the same importance as the information used to design and construct the building. It is produced for the client organisation to use in managing the In Use stage of the building, up to and including its demolition. It provides the basis for every future alteration or change.

The 'As-constructed' Information should be delivered in a format that can be both interrogated and updated by the client or the FM team. It needs to be interoperable with their management systems and software, to accommodate changing operation and maintenance (O&M) requirements and any remodelling, new works and extensions.

Should the 'As-constructed' Information be found to be unreliable or inappropriate then additional surveys may become necessary at a later date, and at an extra cost, in order to re-create an accurate representation of the built space and systems.

'As-constructed' Information will record any changes to key structural or services components and ensure that any subsequent performance implications are reviewed and accepted by the relevant designers, prior to the client accepting the constructed project.

While the content of the 'As-constructed' Information will be defined within the project requirements by the client, operator or FM team throughout the design and assembly stages, the most important requirement is its completeness and accuracy. These requirements can be defined, for example, by referring to:

I dimensional accuracy
Information should be delivered to within a +/- variation of the actual constructed measurement over, say, a particular length and/or height and within a set distance.

I area accuracy
Model file areas scheduled within the Project Information to be within a defined tolerance of the constructed space.

I level of definition (LOD)
Information provided to a defined level of detail and level of information for all components, including any specification and system changes.

I performance accuracy
Information defining environmental or structural performance that reflects the commissioned data within a defined tolerance.

The production of 'As-constructed' Information is frequently left until the later stages of the build and is often compromised by a compressed Construction Programme. The client and FM team should ensure that a validation and verification exercise is rigorously followed prior to accepting the final issue of information to ensure that it meets all the requirements originally set out and accurately reflects the construction.

The 'as-constructed' data requirements of the FM team should also be clearly outlined. They might include:

I the programme for the delivery of the information following completion
I setting-out and tolerance accuracy (including areas and dimensions)
I the LOD delivered for each component and system
I positional and sizing accuracy of the equipment and systems, eg are the ceiling services shown as-constructed or indicatively within the model files
I the level of performance data required within the information (BIM)

I the level of information required in an operational Asset Information Model (AIM)

I the format the information is provided in.

The required level of completeness and accuracy should be outlined within the tender information, the relevant Schedules of Services and the appropriate Project Strategies prior to the Construction stage. How the information will be verified and who will be responsible for its completion should also be set out.

## What are the Key Support Tasks and Project Strategies?

At this stage the Suggested Key Support Tasks developing the Core Objectives and the Information Exchange include reviewing and updating the:

I Sustainability Strategy

I Handover Strategy

I Construction Strategy

I Health and Safety Strategy

I Project Execution Plan, including the Technology and Communication Strategies.

The Project Execution Plan will include summaries of these strategies, which should be reviewed regularly throughout the stage. The 'As-constructed' Information requirements will be developed on the basis of these reviews and the project requirements outlined the Technology and Communication Strategies and protocols.

## Completing 'As-constructed' Information

The operational phase of a project, including its life cycle and in-use costs, is becoming more important in the context of sustainability and carbon reduction. The value of 'As-constructed' Information needs to be seriously considered by the client and the rest of the project team at the outset, as poor information will hinder any improvements in performance and the future development and use of the building.

In some instances the client may be prepared to accept the last construction issue of information from the design team as the 'As-constructed' Information. In many cases this will be to accommodate time constraints in the programme or to keep costs down, either at the end of the stage or throughout construction, to monitor and update the Project Information. It may also highlight that the 'as-constructed' requirements were not clearly identified at the outset within the appropriate Schedules of Services.

Information delivered in this format needs to be carefully used during the In Use stage, although in the majority of instances the difference between a full site survey and the last construction issue will be negligible. The client needs to confirm the level of accuracy required and request the "As-Constructed" Information in the various contract documents accordingly.

### 'As-constructed' Information

'As-constructed' Information can be used to record:

- that the design has been constructed as specified
- that the area parameters and metrics have been delivered
- accurate data in support of the health and safety file
- details of all key structural elements
- details of all key and hidden services
- the locations of key assets
- information to provide a basis for operation, maintenance and future works.

The introduction of BIM and the latest surveying and scanning technologies will make recording and documenting 'As-constructed' Information considerably more efficient. However, the project team will need to define very clearly whether the design information will be updated or whether it will supplement 'as-constructed' 2D/3D survey models and data.

## Point cloud information

Stage 1 highlighted how point cloud surveys can be used to measure and quantify Site Information to assist the project design. This technology can also be utilised to measure construction progress, allowing any discrepancies from tolerances and in workmanship to be understood and either adjusted or accommodated within the Project Information. In addition, as the construction is completed the building can be surveyed to create an 'as-constructed' geometrical model or survey file that can quickly be compared with the design model files or the federated BIM. This can then be used to adjust the Project Information to provide an accurate digital record of the building, or as the 'as-constructed' model itself.

### Who produces the 'as-constructed' model/drawings?

The client will require 'As-constructed' Information to be prepared either during the construction process or when the construction is complete. As a minimum, a requirement for the contractor to 'red pen' changes to the final construction issue drawings on site should be set out within the tender documentation; it should not be assumed to be part of the contractor's 'standard' services. Subject to being a requirement of their professional services agreements, this information can be used by the design team members to update the Project Information, in either a CAD or BIM format.

## Red pen mark-ups

The final construction issue set of information retained on site can be annotated by the contractor as and when changes occur, so that an 'as-constructed' record is developed with the construction itself. These annotations should include or reference any amendments to the setting out, including revised dimensions, specifications, health and safety implications and O&M requirements. Subject to requirements in their professional services contracts, design team members can then review these mark-ups on a regular basis and decide what needs to be incorporated into the Project Information to form the 'As-constructed' Information, although this is not a standard service.

The Construction stage information requirements should be included in the Building Contract and recorded within the Technology and Communication Strategies, highlighting the expected method, process and standards to be used. These methods can be supported with a final 'as-constructed' survey or, if required, specific zones can be re-surveyed at particular stages during the construction, eg at the completion of the structure, the installation of services to be concealed or a spatial survey of completed rooms. This information can then be updated digitally to reflect any changes to the Stage 4 design information.

### How should the content of 'As-constructed' Information be maintained?

While the contractor will be responsible for recording changes and discrepancies on a record set of information, the Schedules of Services of the design team members who monitor construction on site should highlight who will be responsible for updating the construction information to an 'as-constructed' status. A design team member's time on site can become compressed and some appointments only cover minimum monitoring and witnessing requirements and as such many elements of the construction will not be observed by specific disciplines.

### 'As-constructed' Information formats

It is beneficial to request that the contractor's record set is appropriately stamped, signed and dated as an accurate reflection of what was built. This set can then be referenced in the final 'As-constructed' Information by each specific discipline, highlighting who was responsible for the information reflected on the revised drawing.

Should any design team member be required to produce 'As-constructed' Information based on a contractor's marked-up record set then care should be taken to highlight that it is based solely on information provided by the contractor, and that the accuracy of the information has been verified. If, however, the contractor is given responsibility for continually updating the construction information and issuing revisions on a regular basis and the design team members are resourced to review and comment on the updates then there will be no need for such caveats.

## What information is required for 'as-constructed' drawings?

Primarily 'As-constructed' Information will comprise model data, drawings, documents and schedules issued for construction, updated to reflect the 'as-constructed' status of key elements, components and MEP and structural systems. In addition, the information will include changes made as a result of testing and commissioning and any adjustments made during the final inspection process.

It is important that the drawn information is updated to reflect these changes and not annotated and cross-referenced to site sketches and change documentation. BS 1192:2007 notes that specifications and dimensions recorded on the Project Information should be updated, checked and verified, all previous revisions should be removed and the status code amended to AB (as built) to highlight its status.

### 'As-constructed' Information – checklist

'As-constructed' Information will:

- be reviewed and verified for accuracy and completeness
- contain no references to 'equal to' or 'similar to'
- represent the installed specification
- reflect the construction accurately
- provide all information on contractor-designed systems
- include performance changes
- include any significant residual health and safety risks
- include revisions to maintenance and operational requirements
- include all manufacturers', suppliers' and subcontractors' assembly and component drawings
- be amended to an AB status, with all other revisions removed
- include a digital copy of all information.

In addition, using survey or point cloud data to record the 'built' project will enable spatial and geometrical information to be overlaid on the digital model file or within the Project Information Model (PIM) and adjustments made to the sizes and shapes of spaces and the locations of structures and services, providing a very accurate representation of the constructed project.

'As-constructed' Information will provide value to the client and their FM team if it forms an accurate representation of the facility and its operational systems. Having this information in a digital format or within a model means it can be integrated with the FM software. This will maximise the efficiency of the maintenance and operational teams; for example, response times, maintenance scheduling and environmental controls will be improved, which in turn will benefit overall operating costs and energy performance.

## What format should the 'As-constructed' Information be produced in?

'As-constructed' Information produced using CAD, with model files, drawings, documents, specifications and performance requirements, should be updated and validated using the standards and protocols set out within the Project Strategies. The updated information will comprise CAD files in an interoperable format, such as DWG, which can be edited to incorporate any future changes. The information should also be provided as a read-only duplicate, in PDF or DWF format, so that it can be viewed and printed using most standard operating software and systems. Any digital data that has been referenced into the model should be 'bound' or copied live into the file so that the information is always present, irrespective of where the original is filed or copied to. All the drawings, whether produced in an editable or non-editable format, should include the status AB and a completed SHE (safety, health and environment) box.

Providing DWGs in an editable format may require some additional work to remove layers or levels containing data that is unnecessary for the subsequent stages or that might confuse interpretation when the file is viewed at a later date.

With a BIM approach, in addition to producing the updated drawing files in PDF format, any amended data and geometry should be included within a native 'as-constructed' model file from each specific discipline or within a similarly formatted federated model. The model files may also be required in an interoperable format, such as IFC. Consideration should be given to whether these models are provided individually or in a federated format to help minimise the file size after translation.

Federated models may become very large and difficult to open and operate in an IFC format and so care should be taken to ensure that only the necessary data is included. As the amount of data that can be added and updated within a BIM model is considerable, it needs to be reviewed during Stage 1 and defined in each party's Schedule of Services. If the data has not been added as construction progressed, it can be a daunting task to complete this at the end of this stage.

As the interoperability of BIM software and data management tools develop, it is likely that high-quality data recorded within the 'As-constructed' Information will negate the need for O&M manuals. However, in the interim, 'As-constructed' Information should be produced in a format that can be utilised and integrated within the contractor's O&M documents.

It is best practice for the project to retain this information as a live document, maintained in a digital format, either web based, stored in a secure online site, or on a CD/DVD. Using an online storage service means that the 'as-constructed' data can be reviewed remotely, ensuring unlimited access from any location for the authorised team members. It is important to have at least one complete set of 'As-constructed' Information available at all times, and the client organisation should request a record set on CD/DVD to act as a back-up copy in addition to any set of web-based manuals.

## Chapter summary · 5

Stage 5: Construction will see the completion of the project on site. However, the Project Information will still require further development for use within the operational phase. It is important that the Stage 5 Schedules of Services properly frame the need for the project team to appropriately monitor and review any developments and changes made by the contractor. Whether this includes the resolution of Design Queries, and whether these will be incorporated within the Project Information, needs to be clear.

Updating the Project Information is essential to providing an accurate reflection of the installed systems, the spatial setting out and the overall performance of the building. This 'As-constructed' Information can only be used effectively during the Handover and In Use stages of the project if it is accurate and properly structured.

The Stage 5 Information Exchange is:

I 'As-constructed' Information.

Additional information reviewed might include:

I Design Queries
I Construction Programme
I updated Project Strategies
I Change Control Procedures
I Sustainability Aspirations
I Risk Assessments
I Digital Plan of Work.

# Stage 6

# Handover and Close Out

# Chapter overview

Over this initial period of occupation and use, the performance of many aspects of the facility will be monitored to allow for the fine-tuning of any systems and operational processes arising from the construction. At the end of the stage, all aspects of the Building Contract will have been completed: the defects period will have been concluded and the Project Information will have been updated to reflect any amendments made during this close out period to achieve the contractual performance requirements. Incorporating these changes will be essential for allowing the facilities management (FM) and operating teams to measure and understand the project's performance throughout the In Use stage.

**The key coverage in this chapter is as follows:**

Defining updated 'As-constructed' Information

What are the Information Exchanges at the end of Stage 6?

What is updated 'As-constructed' Information?

Determining the format for updated 'As-constructed' Information

Using standards and protocols to manage updated Project Information

Finalising the Project Strategies

# Introduction

Following the completion of the Construction stage, the project team will be focused on facilitating the successful handover of the project. However, it is also important to ensure that the appropriate time is allocated for updating the 'As-constructed' Information to reflect the resolution of any defects, seasonal commissioning, future operational and maintenance (O&M) requirements and the finalisation of the Handover Strategy.

Furthermore, Stage 6 and Stage 7 activities may see a new project team appointed, with the existing team concluding contractual issues and the new team managing any performance and operational issues arising. It is important that the Technology and Communication Strategies highlight how the Project Information will be updated and exchanged in these instances, to avoid the duplication or loss of data.

The data exchanged during and at the end of this stage will underpin the efficient day-to-day use, operation and management of the completed facility, ensuring that the contract information is complete, the performance achieves the design intent and the required Project Outcomes have been delivered. Primarily the information will include any modifications and changes made to improve the performance of the building systems implemented throughout this stage and will be developed with the input of the building operator or FM team to produce data in a compatible, reusable format.

Updating the Project Information with environmental and performance data gathered during the Handover and Close Out stage will ensure that any future improvements will be based on accurate and verified operational information.

## What are the Core Objectives of this stage?

The Core Objectives of the RIBA Plan of Work 2013 at Stage 6 are:

| Tasks ▼ | **6** Handover and Close Out |
|---|---|
| Core Objectives | Handover of building and conclusion of **Building Contract**. |

During this stage the Core Objectives facilitate the handover of the project and the conclusion of the Building Contract. This process will include the completion of updated 'As-constructed' Information in sufficient detail to provide all the asset information requirements necessary for the operation and maintenance of the project.

From the design team's perspective, this may only include very minor changes to the Information completed at the end of Stage 5, although information regarding the MEP systems, which can be adjusted, reset and sometimes recommissioned over the course of the defects period, will need to be incorporated into the final project data set.

## Defining updated 'As-constructed' Information

Following the completion of the Construction stage and irrespective of the size of the project, it is likely that the project team and the user will be required to implement some form of facilities management to control the daily operation and the continued maintenance of the project's assets over its lifespan. This process will primarily deal with the performance of physical assets, their contribution to the delivery of any strategic and operational objectives and the control of any impacts on the designed environment.

### Assets

PAS 1192-3:2014 *Specification for information management for the operational phase of assets using building information modelling* defines an asset as an 'item, thing or entity that has potential or actual value to an organization'. It highlights that:

– An asset may be fixed, mobile or movable. It may be an individual item of plant, a system of connected equipment, a space within a structure, a piece of land, or an entire piece of infrastructure or an entire building or portfolio of assets.
– Value can be tangible, intangible, financial or non-financial, but in the context of this PAS assets are predominantly physical entities such as systems, equipment, inventory or properties.
– The value of an asset might vary throughout its life and an asset might still have value at the end of its life.

## What are the Information Exchanges at the end of Stage 6?

The Information Exchange required at the Handover and Close Out stage comprises:

I   updated 'As-constructed' Information.

For projects completed in compliance with the UK government's Digital Plan of Work (dPOW), this stage also requires a key Information Exchange:

I   UK Government Information Exchanges: Data Drop 4.

## What information is required for the UK Government Information Exchange

This Information Exchange is identified within the UK government's dPOW as Data Drop 4 and encompasses all the maintenance and operational information required to properly use the finished facility. The data exchanged will, for example, comprise the operational and functional information supplied by the product manufacturers, times to first maintenance, replacement costs and warranty information.

### Information requirements at Data Drop 4

The information requirements for this and other stages are outlined within:

– PAS 1192-3:2014 *Specification for information management for the operational phase of assets using building information modelling.*
– BS 1192-4:2014 *Collaborative production of information. Fulfilling employer's information exchange requirements using COBie. Code of practice.*

At this stage the Project Information Model (PIM) will spatially represent the 'as-constructed' facility and will contain all the information provided by the various contractors to maintain the installed systems and equipment. This information will provide the basis for the management of assets and facilities as identified within PAS 1192-3.

The plain language questions relating to the information produced at Stage 6 also outline a number of issues and areas that should be considered at this stage. These include:

I testing and commissioning information
I performance simulations
I functionality assessments
I life cycle carbon assessments
I life cycle cost assessments
I 'as-built' model accuracy
I commissioning information
I updated O&M manuals, to reflect amended performance and settings

| health and safety attribute information
| an asset replacement plan.

These requirements, irrespective of procurement strategy, can be used indicatively, as a checklist defining the information to be produced during the Handover and Close Out Stage.

## What is updated 'As-constructed' Information?

Following the commissioning period during Stage 5, the completed project will experience a period of use and monitoring that allows all its systems to be adjusted to achieve their optimum performance. This period, typically one year (subject to stipulations within the Building Contract), allows the FM team, operator and user, with support from the project team, to understand how the principal operational processes, systems and controls outlined within the Handover Strategy affect the day-to-day running of the facility. These will most likely include any implications arising through seasonal variations and will allow the project performance and operation to be finely tuned to achieve the designed and required outcomes.

The full operational phase of the facility's assets will commence at handover, and following the completion of the Building Contract, at the end of Stage 6, the Project Information will need to be updated to include any modifications and changes that have been implemented to maintain the performance of the facility and its systems. This information will typically be updated within the 'as-constructed' CAD model files or within the PIM. However, it may be that the project creates an Asset Information Model (AIM) during this stage, which depicts the key systems within an 'information-light' spatial geometry, to integrate within the FM team's own computer-aided facilities management (CAFM) processes.

The Schedules of Services' requirements for updated 'As-constructed' Information should, in addition to incorporating any Stage 6 modifications and changes, outline the appropriate level of definition required and the intended use of the contained information. The Project Information can then be adjusted to reflect these requirements. This can mean that the final 'In Use' model may have both its graphical and non-graphical data significantly reduced in order to accommodate any specific operational requirements or management systems that may utilise the information during the In Use stage of the project. These processes will need to be

## Computer-aided facilities management

CAFM systems comprise a number of technologies and information sources. These can include:

- object-oriented database systems
- CAD systems
- BIM
- proprietary building management system (BMS) interfaces.

These systems will typically have a web-based interface and can be implemented remotely by the FM team, to provide analysis and performance information on specific areas in relation to many simple day-to-day tasks. Data can also be automatically collected from a variety of systems or be updated manually to track, analyse and, where necessary, adjust the performance of the building, such as switching the cooling to a summer setting on unseasonably hot days to ensure that usable areas remain comfortable.

BIM has the potential allow a significant amount of data and information to be presented at handover, providing fast access to O&M processes. Linking this information with CAFM systems will ensure that the practice of operating and maintaining a facility becomes extremely responsive and efficient.

For more details on CAFM refer to:

www.bifm.org.uk/bifm/home
www.wbdg.org/om/cafm.php

determined as early as possible within the project to ensure sufficient time is programmed for the development of an appropriate model during this stage.

## Determining the format for updated 'As-constructed' Information

Understanding the way a project and the information describing it will be managed during the In Use stage will help ensure that the client, operator and FM requirements are incorporated appropriately within the most up-to-date information. The format of the 'As-constructed'

Information should also allow for it to be easily updated, not only during this stage, following the completion of the defects period and any seasonal commissioning, but also throughout the remainder of the life cycle of the project.

Typically, the information produced during the Handover and Close Out stage can be supplied in a number of formats, including:

I  updated 'as-constructed' model files (DWGs and DXFs)
I  updated 'as-constructed' discipline-specific PIMs
I  an updated 'as-constructed' federated PIM
I  an AIM
I  printed drawings (PDFs)
I  documents
I  manuals
I  spreadsheets including COBie (Construction Operations Building Information Exchange) data.

The format of the model files, whether a series of CAD model files, PIMs or an AIM, and any drawings and data sheets should be agreed prior to completion of the 'As-constructed' Information at the end of Stage 5 and recorded within the Technology and Communication Strategies. The

### Asset Information Model

The AIM is set out within PAS 1192-3 as 'data and information that relates to assets, to a level required to support an organization's asset management system'. It also notes that the AIM:

– can relate to a single asset, a system of assets or the entire asset portfolio of an organisation.
– consists of graphical, non-graphical and documentation components as well as metadata.
– is the product of the common data environment.

In addition, the PAS notes that the AIM should be used to:

– manage all information regarding works to an asset
– link information about works to assets
– receive information throughout the project stages, up to acceptance of the 'as-built' PIM.

required level of definition, the end-user software and the native software capabilities should also all be reviewed prior to deciding how the final information will be formatted and communicated. This format can then be progressed during Stage 6 to develop the In Use information.

FM requirements outlining the safe operation and maintenance of the project will principally be recorded within the Stage 5 'As-constructed' Information and the manufacturers' installation, specification, health and safety, performance and maintenance and operational data sheets within the O&M manuals. These CAD model files and supporting documents can be maintained as 'live' information and updated as necessary to reflect any changes that occur during this and subsequent stages.

Alternatively, if an 'as-constructed' PIM has been progressed with digital links to supporting documents then these or an AIM can be used to receive updated 'As-constructed' Information.

The 'as-constructed' PIM can be developed and utilised as an AIM during this stage or at the start of Stage 7: In Use. Alternatively, a separate AIM can be produced from the 'as-constructed' model comprising simplified geometry, defining only the project context but with enough complexity to depict the key assets and systems to be operated and maintained.

The contents of the AIM will typically be set out within the Asset Information Requirements (AIRs), both of which are reviewed further in Stage 7: In Use (see page 214).

Future updates to any operational systems, the physical structure or the building form might also be developed as separate model files. These will be updated, federated and coordinated with the Project Information until the data is fully verified, when it can then be integrated within the most current PIM.

### How should updated 'As-constructed' Information be delivered?

Digital copies of 2D information can be managed, collated and exchanged relatively easily, either via a web-based portal or on CD or DVD. The success of an information model, however, will rely on the exchange standards within each discipline's native application to ensure that the separate models can be federated and viewed by the operator, owner or FM team. A more common approach is to use an interchangeable file format, which will ensure that all the valid information can be accessed, understood and used once recorded.

A PIM or an AIM can represent all the validated 'As-constructed' Information in a single federated model. Both will ideally be created in an interchangeable format, such as an IFC (Industry Foundation Class) file. IFC is primarily a read-only file format, so it is not possible to incorporate any future changes or performance analysis into an IFC file unless it has been imported directly into compatible authoring software. While this format is good for maintaining a copy of the exchanged data, updating the model file with new information will require a revised IFC model to be created following the translation and modification of the original in appropriate software. Because of these multiple translations, the validity of the information should always be checked and verified, following each iteration, to ensure that corruption and data loss are minimised.

Maintaining information model files such as PIMs and AIMs can require a level of expertise that may not exist within a client organisation or FM team. Therefore, it might be necessary to simplify the 'as-constructed' model to depict only simple spatial geometries, so that it can be supported by the FM management software and systems. These 'information light' models can be supplemented with a series of CAD model files and spreadsheet data, such as COBie, to provide an interface with the FM software.

### 'Information light' models

A BIM model will typically comprise objects modelled to a particular level of definition combined with a number of instances of data to describe the object, how it is procured and how it is constructed, operated and maintained. While this information supports the design and construction of the project, during Stages 6 and 7 it may make some model files unnecessarily large and difficult to use within an FM environment. To overcome these issues the modelled elements (geometry) can be simplified and saved to new files. These files will contain only the minimum information needed to allow FM and other operational software to have an understanding of the spatial arrangement of the project. Key operational data is then used via other sources (such as spreadsheets) to control the use and operation of the space.

The use of BIM in an FM environment is continually developing. Currently, CAD and modelling approaches are being used together, to accommodate each project team's abilities and requirements. For the foreseeable future it is likely that FM systems will be required to combine formats, until processes and requirements become standardised and more widely used in producing and managing building information.

## Using standards and protocols to manage updated Project Information

Controlling the production and completeness of the Project Information will provide significant benefits within an asset management approach, whether for a single facility or a wide portfolio of projects. Accurate information will improve operational awareness, management and future planning throughout the life cycle process. PAS 1192-3 has been developed to support this type of approach, specifying appropriate processes for the exchange of data to assist the operation, maintenance and strategic management of assets.

### How does PAS 1192-3 help in the management of post-construction information?

PAS 1192-3 highlights the types and nature of data and information required to develop a federated AIM. The document, building on the information standards and processes set out within PAS 1192-2:2013, suggests that the post-construction information requirements are recorded

### Standards and protocols

PAS 1192-3:2014 *Specification for information management for the operational phase of assets using building information modelling* can be downloaded from: http://shop.bsigroup.com/forms/PASs/PAS-1192-3/

In addition to the above, the British Standard BS 1192-4:2014 *Collaborative production of information. Fulfilling employer's information exchange requirements using COBie. Code of practice* highlights how a COBie output can be utilised to demonstrate compliance with the Employer's Information Requirements.

within an AIRs document. The PAS recommends this is completed at the outset of the project and used to inform the Employer's Information Requirements (EIRs) or brief.

The AIRs should outline guidance for:

I an organisation's information management processes, procedures and activities
I how and what information will be exchanged
I any specific roles and responsibilities for information management.

The PAS recognises that facilities management can encompass numerous project-related areas, including finance, quality, strategic overviews, as well as operations, maintenance and performance. However, it highlights that its guidance is only concerned with the management of information associated with the performance of physical assets, as opposed to the more day-to-day operational aspects of a facility.

The approach outlined within PAS 1192-2 and PAS 1192-3 does not necessarily require an organisation to produce information within a BIM model. However, projects required to achieve an information maturity of BIM Level 2 will need to follow the structure and processes outlined within both specifications.

## Finalising the Project Strategies

Following the completion and commissioning of the operational building systems, the Project Strategies should, throughout Stage 6, continue to ensure that any changes made as a result of use, operation and/or maintenance requirements during this period are accommodated within the updated Project Information.

### What are the Stage 6 Key Support Tasks?

The 'As-constructed' Information should be updated to include all of the information gathered throughout the Stage. It is fundamental that the nature and format of the updated 'As-constructed' Information is highlighted within the Project Execution Plan strategies so that the appropriate data can be monitored, collated and updated to inform the In Use stage tasks and objectives.

The Suggested Key support Tasks outlined at this stage include activities related to:

I   the Handover Strategy
I   Feedback
I   updated Project Information.

**How is the Handover Strategy finalised?**

Following the project handover and the completion of the O&M manuals and health and safety file, the 'as-constructed' model files or the PIM or AIM will continue to be updated to reflect any changes made throughout Stage 6. This will ensure that any future monitoring and analysis conducted during the In Use stage is premised on accurate and verified data.

## Handover information – checklist

The Handover Strategy and Maintenance and Operational Strategy will be finalised to include:

– a summary of the building
– building services performance requirements
– any system zoning
– the location of relevant plant and equipment
– make and model numbers of all significant items of plant and equipment
– manufacturers' contact details and instructions for relevant plant and equipment
– a schedule of the energy-consuming services
– a non-technical user guide
– guidance manual on daily, monthly, seasonal and annual maintenance and operational requirements
– 'as-constructed' drawings and schematics of the building and systems
– location of the PIM/AIM and a list of Stage 6 revisions and amendments
– commissioning records, including demonstration of compliance with specified energy standards
– health and safety file, as required by the CDM Regulations.

Handover information – checklist (*continued*)

The Handover Strategy and subsequent Building Contract should, in
order to benefit works completed during Stage 6, highlight that:

– any equipment subject to a manufacturer's/supplier's warranty
  is tendered to allow the specific supplier/installer to provide
  maintenance during the warranty period if not being operated by the
  FM team
– tender and specification documentation includes a provision for
  contractors, subcontractors and suppliers to provide operational
  training for users and the FM team before handover.

## Chapter summary     6

While a Core Objective of the Handover and Close Out stage is the
conclusion of the Building Contract, including the resolution of
any defects and performance issues arising throughout this initial
period of use, the Project Information will be further developed to
reflect the operational status of the project at the end of the stage.

Changes that are made during the handover period may require
some elements and systems to be adjusted and recommissioned.
The revised settings and information will be collected and collated
throughout Stage 6 and updated within the project model files and
documents.

The project protocols and strategies will define how the Project
Information will be developed and who will be responsible for
recording and updating any Feedback, performance and handover
requirements. The client should ensure that this data is accurately
updated, and that it is produced in a format that will allow it to
be developed into an operational data set for use in monitoring,
analysing, operating and maintaining the project throughout the
In Use stage.

The Stage 6 Information Exchange is:

I  updated 'As-constructed' Information.

Additional information reviewed and activities completed might include:

I  completion of the Building Contract
I  concluding Project Strategies
I  Feedback
I  Project Performance Evaluations
I  Post-occupancy Evaluations
I  building performance evaluations
I  Digital Plan of Work.

# In Use

# Chapter overview

Beyond the design and construction stages, the Project Information sets out the basis from which to test and analyse the building in use. This process enables the key areas of success and failure to be understood, highlighting areas of improvement that can be used to inform future designs of a similar size and scope. This information can also be used strategically within Stage 0 of a similar project, ensuring the required Project Outcomes are aligned with known performance and design solutions. More importantly, the spatial and system performance data collected can be utilised to improve the day-to-day use and operational efficiencies of the facility. These cyclical approaches, iteratively developing and assessing measured information and actual outcomes, will, as more data is collected, also offer improvements in cost, quality, operation and value, while preventing a repeat of any issues and errors on similar projects.

**The key coverage in this chapter is as follows:**

Measuring the success of the project

What are the Information Exchanges at the end of Stage 7?

What are the Key Support Tasks and Project Strategies?

Why is Feedback important?

Using Feedback to improve future outcomes

Understanding how Post-occupancy Evaluation and Feedback can benefit similar projects

# Introduction

Stage 7: In Use is a new stage, covering the life and operation of the building after handover through to its eventual reuse or demolition. It includes gathering information on the performance of the completed project, such as the efficiencies of the building systems and spaces, and it allows for assessments, through both Post-occupancy Evaluation (POE) and building performance evaluation (BPE), to determine whether the Project Outcomes and aspirations have been achieved.

The handover duties are completed throughout Stage 6, prior to conclusion of the Building Contract. Traditionally, the project team appointed to undertake the design and construction is rarely involved when a building is actually in use. However, with a shift in focus towards the energy and carbon performance of buildings in use, some activities may now be required beyond Stage 6. The principal tasks will be to monitor actual performance and use and to update the asset information accordingly, and as such additional input from the project team may be required.

## What are the Core Objectives of this stage?

The Core Objectives of the RIBA Plan of Work 2013 at Stage 7 are:

At this stage the Core Objectives for the project team will be project specific and outlined within each team member's Schedule of Services. The Final Project Brief, Design Responsibility Matrix and Schedules of Services may include the requirement for a Soft Landings-type approach, involving the monitoring of systems for a period of years beyond handover, as well as further requirements for recording and interpreting Feedback, POEs and BPEs and performing other tasks and activities that relate to the performance of the project in use.

## Measuring the success of the project

Learning from the end to define the beginning is critical to the structure of the RIBA Plan of Work 2013, allowing the successful and, in some cases, the not so successful objectives, tasks and information to be measured and assessed. Stage 7 finalises a cyclical approach to the use and production of Project Information to inform Research and Development and Feedback requirements outlining the aspirations and performance requirements of subsequent schemes.

## What are the Information Exchanges at the end of Stage 7?

The Information Exchanges required during the In Use stage comprise:

I Feedback
I updated asset information.

Understanding the successful operation and use of a building in relation to the required Project Outcomes, energy performance and life cycle costs will provide essential feedback for the project team. This can be used to improve the day-to-day operational efficiencies of a particular project, but it will also highlight valuable opportunities to improve the process and design of future projects.

For projects completed in accordance with the UK government's Digital Plan of Work (dPOW), this stage also requires an Information Exchange aligned with the requirements of PAS 1192-3:2014:

I UK Government Information Exchanges: Data Drop 5 and as required thereafter.

The requirements for Data Drop 5 are not definitively set out as yet but will most likely cover project-specific issues. However, the concept of using feedback and reviews to understand how assets are operated and managed is considered an essential activity and forms an integral part of the Government Soft Landings (GSL) initiative (a similar approach to the BSRIA Soft Landings process outlined in Stage 0 but with a greater focus on the facilities management (FM) aspect of occupation and use). The information at this stage will contain all the base operational data.

However, PAS 1192-2:2103 and PAS 1192-3 suggest that this might be developed further as necessary, ensuring that the information develops with the life of the building and is always up to date.

As with Stage 6, the plain language questions (PLQs) outline a number of areas that 'in use' information should address at this stage, including:

I model management and information updates
I building performance evaluations
I Post-occupancy Evaluations
I energy assessments
I water usage
I metered performance
I long-term aftercare
I decommissioning information
I recyclability and safety information.

These requirements can be used indicatively, as a checklist defining the information that might be required to enhance the performance of the building and reduce its energy use and life cycle costs. A full list of this stage's indicative PLQs and detailed explanations can be reviewed at www.thenbs.com/BIMTaskGroupLabs/questions.html.

## What are the Key Support Tasks and Project Strategies?

At this stage, the Suggested Key Support Tasks and Project Strategies, including those set out within the Project Execution Plan, will support continual updates of the Project Information using any data gathered throughout the use of the facility. This information should be recorded in a format that ensures the client can amend the asset and 'As-constructed' Information as required over the lifespan of the project.

Many of the strategies will be progressed by the FM team, and in order to maintain the accuracy of the data, any updates and changes will need to be recorded in accordance with the project protocols and standards. This may involve updating the model to include performance data, or to highlight elements of the model as examples of best practice against which to benchmark particular Project Outcomes in future projects.

The Project Strategies and activities to be concluded in Stage 7 include:

I  Handover Strategy
I  Post-occupancy Evaluation
I  Building Performance Evaluation
I  Project Performance
I  Project Outcomes
I  Research and Development
I  updating Project Information
I  Feedback.

It is likely that the completion of these strategies will impact some members of the project team more than others, with many of the performance reviews being completed by the MEP engineers and the POE and Feedback analysis completed by the architect. Contractors may also carry the responsibility for concluding these Project Strategies, including the updating of Project Information and data associated with any Soft Landings processes that may be required.

## Why is Feedback important?

Measuring and understanding the physical performance of the project, such as cooling performance on a hot day or heating on a cold day, is a relatively simple process: measuring how successful the performance and use of the spaces are is a little more complicated. However, the way users manage systems and respond to changes can be evaluated using targeted surveys, analysis and a series of POE reviews, to give an understanding of the full impacts of the way the building works and is occupied.

Irrespective of the information format (CAD or BIM), the Project Information should be continuously updated to incorporate these forms of feedback. An example would be amendments to the set points on the heating and cooling systems to suit occupational variations throughout the course of the day or week.

The client should ensure that the appropriate resources will be available at this stage to record and update any impacts that these changes may have on the operational and maintenance requirements of each asset. If a BIM approach has been used then the client will need to specify in the Organisational Information Requirements (OIRs) whether the updated federated model or a separate Asset Information Model (AIM) (see AIRs and OIRs, page 214) will be available in a format that will allow its integration with the FM systems and the incorporation of Feedback for the client and future development teams to use. Any information that is added to the operational model will need to be appropriately validated; PAS 1192-3 highlights how the common data environment processes manage the way this information is incorporated into the AIM or Project Information Model (PIM). This follows the principle that updates are classed as work in progress (WIP) until they have been checked and approved by the client or information manager.

Feedback and benchmarking information gathered over a period of two or three years, or perhaps an even longer, will provide extremely useful data for assessing and developing the performance and energy efficiency of the project against other successful projects of a similar size and type, as well as highlighting what might be successful on others.

### Feedback

As a minimum, Feedback reviews should provide answers to the following questions:

- How effective are the ongoing management processes?
- Does the building perform as intended?
- What needs to change?
- Are life cycle costs in line with predictions?
- How can better efficiencies be achieved?
- What problems may need resolution quickly?
- Where can improvements be made?
- Have any of the users' needs changed?
- What can be learnt and applied to future projects?

## Using Feedback to improve future outcomes

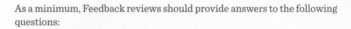

Completed projects should be subjected to an annual review for a number of years after handover. The aim will be to develop data that can be used to improve the efficiency and operation of the completed systems and to help improve the occupancy and use of the spaces. These reviews will assess a range of issues, from occupant satisfaction to energy and environmental controls, improving the overall sustainability of the facility as well as reducing operating costs.

### Feedback reviews

Innovate UK (formerly the Technology Strategy Board), through its Building Performance Evaluation Group, has highlighted a number of tools and templates that can be utilised during these reviews in order to ensure that the feedback is useful. These include guidance documents for conducting studies for:

- domestic projects
- non-domestic projects
- follow-on projects

and also template documents for reporting studies for:

- domestic projects
- non-domestic projects.

These documents, and also a list and links to a number of other monitoring tools, can be found online at: https://connect. innovateuk.org/web/building-performance-evaluation

The templates suggest agendas and questions that might be asked at each review to help improve understanding of how the project is performing and what can be done to improve the way it is utilised and perceived. Areas covered include:

- building operation and usage patterns
- system operation
- building management system
- other controls and interfaces
- energy and water consumption/management.

### How can understanding how the project is used improve its performance?

POE helps the client understand and assess whether the performance of the building has achieved the brief aspirations and Project Outcomes. It can highlight initial issues with use and performance and identify early on any learning requirements for users (eg help with understanding the controls and system operations), while also comparing output performance with that of similar completed facilities.

Reviews can be used to:

I  assess design quality
I  benchmark against best practice examples
I  understand the use of the space
I  assess whether spaces are comfortable
I  establish the productivity and performance of occupiers
I  determine whether user needs and satisfaction have been achieved as planned
I  monitor whether spaces are used as intended
I  monitor how controls and systems impact the use of the space
I  assess how planned adjacencies actually function
I  understand the impact of materials on the use of the building
I  understand how environmental conditions impact the use.

A number of different proprietary systems for POE were identified in Stage 0 (see box on page 18). These can be used for recording and assessing POE requirements to ensure that poor building performance does not unnecessarily impact occupant comfort, running costs or business efficiency. The process of POE ensures that the final Project Information remains focused on practical issues, functionality and achieving the Project Outcomes. As a result, its value is being increasingly appreciated and its use is becoming mandatory on many public projects.

### What are the BIM benefits for Post-occupancy Evaluation?

If a PIM has been developed to incorporate 'as-constructed' and operational data then it can be utilised to help develop and maintain the facility in use. Whilst most of the data will interface with the building management system and software, using the BIM can help visualise the space where any improvements can be made.

The model will need to be updated and the client will need to identify who will perform this task. If the lead designer managed the BIM process it may be they who are retained to make annual updates to the BIM, or the FM team may manage this work themselves.

Keeping the model updated ensures that accurate data is always to hand and that any re-planning, refurbishment or extension projects can use current and up-to-date information, avoiding the need for additional surveys and new modelling.

## Understanding how Post-occupancy Evaluation and Feedback can benefit similar projects

The information gained from POE and Feedback can be used to inform the Project Outcomes and aspirations of a client's next project. Alternatively, it can, with the client's permission, be used as a benchmark, reflecting best practice within both sector and building-type approaches. Using this information will allow future designs and products to develop and respond to the ways that successful projects are designed, constructed and operated.

### What benefits can asset information provide during the In Use stage?

Computer-aided facilities management (CAFM) can be used to organise and manage a building's physical assets throughout the In Use stage. This can be defined in both the OIRs and the Employer's Information Requirements (EIRs) and can comprise either whole systems or individual components, such as doors, windows, pumps or fans.

Using the updated 'As-constructed' Information to inform the FM requirements enables the team to create and monitor asset schedules, asses the performance and utilisation of specific spaces, outline health and safety issues, and monitor many other attributes of the building, such as the:

I ages of assets
I cost data for asset replacement and operation
I life expectancy of assets
I construction data
I contract and warranty data
I building manager's contact details
I technology requirements and updates for system software.

CAFM systems can monitor operational tasks and activities to improve the performance of the assets and the efficiencies of the management team.

This will ultimately improve the overall energy and carbon performance of the project over the life of the facility.

## Structuring information for the life of the project

The Project Information produced during Stage 6 will comprise either a number of 'as-constructed' CAD model files and associated documents or an updated 'as-constructed' PIM. The information utilised by the FM team and operator will be premised on this data and, as outlined within PAS 1192-3, will help to:

I define the asset management strategy
I implement asset management plans
I manage the asset life cycle
I manage asset knowledge
I manage the organisation and its human resources
I manage and review risk.

The Project Information may also be progressed during Stage 6 and again during Stage 7 as a separate AIM, comprising a simplified geometry and the operational, performance and maintenance information required for each asset. The AIM will be updated periodically through Information Exchanges between the FM team or operator and those responsible for its development. The frequency and scope of these Information Exchanges will be project specific and will be defined by the client responsible for setting out the OIRs and AIRs.

### AIRs and OIRs

The information requirements for the In Use stage are defined at the early stages of the project and outlined within the Employer's Information Requirements (EIRs) for new-build projects. Refurbishments and extensions might only use Organisational Information Requirements (OIR) to define the requirements for the works. Figure 7.1 shows the hierarchy of information that defines what should be updated within the Asset Information Model (AIM).

Asset Information Requirements (AIRs) are defined within PAS 1192-3 as the data and information requirements of the

## AIRs and OIRs (*continued*)

organisation in relation to the asset(s) it is responsible for.

- The organisation should have processes in place to respond to any changes in the AIRs while a project using PAS 1192-2 is under way.
- When the asset management process gives rise to a new project that will apply PAS 1192-2, then the appropriate AIRs become or form part of the EIRs, as defined in PAS 1192-2.

OIRs are defined within PAS 1192-3 as the data and information required to achieve the organisation's objectives.

- The management activities leading to OIRs are the equivalent of the employer's key decision points in PAS 1192-2.

For further details, refer to PAS 1192-3:2014, which outlines a number of protocols and examples suggesting how the AIM, AIRs and OIRs can assist the process of updating and managing the In Use Project Information.

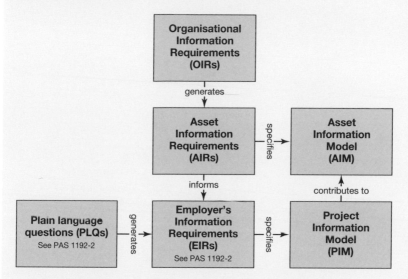

*Figure 7.1   Relationships between the elements of information management in PAS 1192-3:2014*

## How can information be recycled?

Stage 0: Strategic Definition identified that, with the emergence of information modelling systems, it is possible to use successfully modelled representations of a built project to form part of a 3D digital brief for the next project or a similar building type. Using BIM objects and components in this way will allow the design team, manufacturers and suppliers to develop improved systems on the basis of feedback and performance evaluations, to meet specific outcomes and brief requirements.

The models can define the spatial, environmental and construction constraints as well as softer constraints, such as minimum dimensions and clearances required by key operational equipment or by statutory authorities, to demonstrate compliance. In addition, the performance specification and the likely outturn costs can be embedded within the model to add a higher level of detail to the brief, which may help to reduce the risks in early responses to the required Project Outcomes.

### Digital objects

While the government is reviewing the use of modelled zones as briefing documents, suppliers and manufacturers are producing more detailed functional objects and components for use within modelled projects. Many of these can be downloaded from providers such as:

– the NBS National BIM Library: www.nationalbimlibrary.com
– the BIMStore: www.bimstore.co.uk

The workflow suggested by the NBS, as outlined in figure 7.2, ensures that the data utilised can be maintained and informed throughout the development of the project.

Library or benchmarked objects used as part of a 3D brief can include the relevant specification and the required graphical and non-graphical outputs to suit varying levels of definition, allowing them to be used and developed accordingly with the rest of the Project Information.

*Figure 7.2    National BIM Library objects workflow progression*

Each element/object or space will require validation and checking by the design team to ensure that it complies with the relevant design standards and project protocols and, more importantly, is compatible with the Project Outcomes.

## Chapter summary                                    7

As a new stage within the RIBA Plan of Work, Stage 7: In Use allows the client, end user and operator to utilise the knowledge and expertise of the project team to monitor and analyse the performance of systems and spaces for as long as required. These reviews can help the owner and/or user to fine-tune the facility for both seasonal and occupational variations, so that energy, cost and carbon performance are optimised and the facility is both comfortable and satisfying to occupy.

PAS 1192-3 provides guidance on how the information requirements for this stage should be specified, collated and managed. It is supported by a number of industry guidance documents highlighting how data should be reviewed, gathered and assessed.

Investigating how a building performs through Post-occupancy Reviews and Feedback creates a large database of information, which can inform both Research and Development and Feedback requirements at the Strategic Design stage on similar projects. More importantly, In Use information allows the impact of the Project Outcomes and aspirations to be accurately assessed, providing the basis for improving the quality and performance of the building itself and future designs.

The Stage 7 Information Exchanges are:

I  asset information
I  Feedback.

Additional information and activities completed might include:

I  concluding Project Strategies
I  Project Performance evaluations
I  Post-occupancy Evaluations
I  building performance evaluations
I  Research and Development
I  Digital Plan of Work.

# Further reading

## Relevant standards

BS 1192:2007          *Collaborative production of architectural,
                       engineering and construction information*

PAS1192-2:2013        *Specification for information management
                       for the capital/delivery phase of construction
                       projects using building information modelling*

PAS 1192-3:2014       *Specification for information management
                       for the operational phase of assets using
                       building information modelling*

BS 1192-4:2014        *Collaborative production of information.
                       Fulfilling employer's information exchange
                       requirements using COBie. Code of practice*

BS 8541-3:2012        *Library objects for architecture, engineering
                       and construction. Shape and measurement.
                       Code of practice*

BS EN ISO 14040:2006  *Environmental management. Life cycle
                       assessment. Principles and framework*

BS ISO 12006-3:2007   *Building construction. Organization of
                       information about construction works.
                       Framework for object-oriented information*

BS ISO 16739:2013     *Industry Foundation Classes (IFC) for data
                       sharing in the construction and facility
                       management industries*

BS ISO 29481-1:2010   *Building information modelling. Information
                       delivery manual. Methodology and format*

## Books and reports

### RIBA Plan of Work 2013 Guides:

| *Contract Administration*
  Ian Davies, RIBA Publishing, ISBN: 9781859465523
| *Design Management*
  Dale Sinclair, RIBA Publishing, ISBN: 9781859465509
| *Project Leadership*
  Nick Willars, RIBA Publishing, ISBN: 9781859465516
| *Town Planning*
  Ruth Reed, RIBA Publishing, ISBN: 9781859465530

*Assembling a Collaborative Project Team: Practical Tools Including Multidisciplinary Schedules of Services*
Dale Sinclair, 2013, RIBA Publishing, ISBN: 9781859464977

*BIM: Management for Value, Cost and Carbon Improvement. A Report for the Government Construction Client Group*
Building Information (BIM) Working Party Strategy Paper, July 2011, Department for Business, Innovation and Skills

*BIM in Small Practices: Illustrated Case Studies*
Robert Klaschka, 2014, RIBA Publishing, ISBN: 9781859464991

*Building Information Management: A Standard Framework and Guide to BS 1192*
Mervyn Richards, 2010, British Standards Institution,
ISBN: 9780580708701

*COBie Data Drops: Structure, Uses and Examples*
BIM Task Group, 2012

*Government Construction Strategy*
Cabinet Office, May 2011

*Government Construction Strategy: One Year On Report and Action Plan Update*
Cabinet Office, July 2012

*Guide to Using the RIBA Plan of Work 2013*
Dale Sinclair, 2013, RIBA Publishing, ISBN: 9781859465042

*Putting You in Control: The RIBA Client Design Advisor*
RIBA, 2009, www.architecture.com/Files/RIBAProfessionalServices/
    Directories/ClientDesignAdvisor.pdf

*RIBA Job Book* (ninth edition)
Nigel Ostime, 2013, RIBA Publishing, ISBN: 9781859464960

*Small Projects Handbook*
Nigel Ostime, 2014, RIBA Publishing, ISBN: 9781859465493

*The Value Handbook*
Commission for Architecture and the Built Environment, 2006, CABE
(Design Council), ISBN: 1846330122

**BSRIA Soft Landings:**

| | |
|---|---|
| BSRIA BG 4/2009 | *The Soft Landings Framework: For Better Briefing, Design, Handover and Building Performance In-use* |
| BSRIA BG 38/2012 | *Soft Landings Core Principles* |
| BSRIA BG 45/2014 | *How to Procure Soft Landings – Guidance* |

## Protocols and specifications

*Building Information Modelling (BIM) Protocol (CIC/BIM Pro):*
*Standard Protocol for use in Projects using Building Information*
*Modelling*
Construction Industry Council, 2013

**AEC (UK) CAD and BIM standards:**
AEC (UK) CAD Standards For Layer Naming v3.0
AEC (UK) Model File Naming Handbook v2.4
AEC (UK) Drawing Management Handbook v2.4
AEC (UK) BIM Protocol v2.0 (Main document)
AEC (UK) BIM Protocol – BIM Execution Plan v2.0
AEC (UK) BIM Protocol – Model Matrix v2.0
AEC (UK) BIM Protocol Model Validation Checklists for Autodesk,
Bentley and Graphisoft systems

*AIA Document G202–2013: Project Building Information Modeling*
*Protocol Form*
American Institute of Architects, 2013

*2014 Level of Development Specification*
BIM Forum, 2014

## Key websites

### Benchmarking
www.ajbuildingslibrary.co.uk
www.architecture.com/RIBA/Visitus/Library
www.bco.org.uk/Research/Best-Practice-Guides.aspx
www.breeam.org/case-studies.jsp
www.building.co.uk/buildings/technical-case-studies
www.building.co.uk/data/cost-data/cost-model
www.carbonbuzz.org
www.carbontrust.com/resources/faqs/sector-specific-advice/energy-
    benchmarking
www.cibse.org/building-services/building-services-case-studies
www.rics.org/uk/knowledge/bcis

### British Standards and protocols
shop.bsigroup.com/

### Building Information Modelling standards and protocols
www.aecuk.wordpress.com
http://bimforum.org
www.bimtaskgroup.org
www.buildingsmart.org.uk
www.cic.org.uk/publications
www.cpic.org.uk

### Building performance evaluation
www.connect.innovateuk.org/web/building-performance-evaluation
www.instituteforsustainability.co.uk/guidetobpe

### COBie
www.bimtaskgroup.org/cobie-data-drops/

### Computer-aided facilities management
www.bifm.org.uk/bifm/home
www.wbdg.org/om/cafm.php

**Digital Plan of Work Toolkit**
https://toolkit.thenbs.com

**Plain language questions**
www.thenbs.com/BIMTaskGroupLabs/questions.html

**Planning advice**
www.planningportal.gov.uk

**Post-occupancy Evaluation**
www.bre.co.uk
www.connect.innovateuk.org/web/building-performance-evaluation
www.dqi.org.uk

**RIBA Client Design Advisor**
www.architecture.com/Files/RIBAProfessionalServices/Directories/
       ClientDesignAdvisor.pdf

**RIBA Plan of Work 2013 Toolbox**
http://www.ribaplanofwork.com

**Soft Landings**
www.bsria.co.uk/services/design/soft-landings/free-guidance
www.usablebuildings.co.uk

**Uniclass**
www.cpic.org.uk
www.cpic.org.uk/uniclass2
www.thenbs.com/uniclass/

# Information exchanges glossary

## Asset Information Model (AIM)

Model developed during Stages 5 to 7 to include all as-installed components and systems' operational data to support the management and operation of the facility in use.

## Asset Information Requirements (AIRs)

Set out the post-construction information requirements to be delivered within the AIM and any other data record, such as COBie.

## Building Information Modelling (BIM) dimensions

- 3D: Model
- 4D: Time
- 5D: Cost
- 6D: Facilities management

## BIM execution plan (BEP)

Plan produced to outline the project protocols, standards and procedures for managing the delivery of the Project Information.

## CAD

Computer-aided design/drawing.

## Common data environment (CDE)

An information management system and process forming part of the Standard Method and Procedure outlined within BS 1192:2007. More commonly defined as a single source of information used to collate and manage project documentation, such as a web-based system or a project extranet.

## CI/SfB

Construction Industry/Samarbetskomitten for Byggnadsfragor; a construction classification system comprising a combined UK and Swedish approach. The system has been in use for a number of years and is being replaced by Uniclass, but CI/SfB can still be found in use for organising collections in many libraries.

## Construction Operations Building Information Exchange (COBie)

Identifies the non-geometric information that is needed to exchange managed asset information over the life of a project, which is typically presented in a spreadsheet format.

## Data

Unorganised facts stored but not yet interpreted or analysed

## Data Drop

The release of defined information in accordance with the development of the project as set out within the Employer's Information Requirements (EIRs). Compliance with a BIM maturity of Level 2 requires all Data Drops to be in accordance with COBie standards.

## Digital Plan of Work (dPOW)

Sets out the project workflow and requirements as outlined in a number of UK government BIM standards and protocols. The concept has been further developed by the NBS within a free-to-use web-based BIM Toolkit, which enables the project to 'define the team, responsibilities and an information delivery plan for each stage of a project'.

## Document management/electronic document management system (DM/EDM)

A system that provides for the storage,

sharing and management of project-related information, documents and files, electronically or otherwise, with a greater level of security and accessibility than a typical standard operating system.

## DWF

CAD file format: allows the same level of interrogation of the information as a DWG file, but files are more compressed and typically smaller. This format is not editable.

## DWG

CAD file format: can allow any competent CAD operator/technician to measure and review and also make modifications and updates to reflect any changes made during the handover period and in-use phase.

## Employer's Information Requirements (EIRs)

Set out what information the client needs to ensure that the design is developed appropriately for the construction and operation of the completed project. This document will accompany the Project Brief.

## Federated model

A single project model composed of a number of separate, unlinked discipline-specific Project Information Models, such as the architectural, structural and MEP models.

## IFC (Industry Foundation Class)

An object-based file format, used to enable the exchange of information between different software platforms. IFC is an official standard and recommended for Level 2 BIM compliance.

## Information

Data that has been structured, organised or presented in a given context to make it suitable for communication, interpretation or processing.

## Level of definition (LOD)

Defined in PAS 1192-2:2013 as the collective term for the 'level of model detail' (LoD) and the 'level of information detail' (LoI). LoD sets out the graphical content within a system or component and LoI sets out the information required to describe and maintain it at each stage of the project. LOD fundamentally outlines the reliability of the Project Information, highlighting what it can and can't be used for by other members of the project team and stakeholders.

## Model

BS 1192:2007 defines a model as 'a collection of containers organized to represent the physical parts of objects, for example a building or a mechanical device'. It states that 'models can be two-dimensional (2D) or three-dimensional (3D), and can include graphical as well as non-graphical content'.

## Model Production and Delivery Table (MPDT)

A type of Design Responsibility Matrix outlined within the CIC's BIM Protocol appendices; it allows the LOD to be defined at each specific project stage.

## Organisational Information Requirements (OIRs)

Defined within PAS 1192-3:2014 as the data and information required to achieve the organisation's objectives.

## PDF

A file format for exchanging documents. PDF documents have a relatively small file size but cannot be updated.

**Plain Language Questions (PLQs)**

Set out what Information is required at the end of each stage, informing the EIRs and validating the information produced to ensure the project can progress.

**Project Information Model (PIM)**

Described by PAS 1192-2:2013 as the 'information model developed during the design and construction phase of a project'. A PIM is likely to include a number of individual discipline's models or be a federated model that includes all non-graphical data and supplementary documentation.

**Standard Method and Procedure (SMP)**

Eight key areas for the management of project information, as set out within BS 1192:2007.

# RIBA Plan of Work 2013 glossary

A number of new themes and subject matters have been included in the RIBA Plan of Work 2013. The following presents a glossary of all of the capitalised terms that are used throughout the RIBA Plan of Work 2013. Defining certain terms has been necessary to clarify the intent of a term, to provide additional insight into the purpose of certain terms and to ensure consistency in the interpretation of the RIBA Plan of Work 2013.

### 'As-constructed' Information

Information produced at the end of a project to represent what has been constructed. This will comprise a mixture of 'as-built' information from specialist subcontractors and the 'final construction issue' from design team members. Clients may also wish to undertake 'as-built' surveys using new surveying technologies to bring a further degree of accuracy to this information.

### Building Contract

The contract between the client and the contractor for the construction of the project. In some instances, the **Building Contract** may contain design duties for specialist subcontractors and/or design team members. On some projects, more than one Building Contract may be required; for example, one for shell and core works and another for furniture, fitting and equipment aspects.

### Building Information Modelling (BIM)

BIM is widely used as the acronym for 'Building Information Modelling', which is commonly defined (using the Construction Project Information Committee (CPIC) definition) as: 'digital representation of physical and functional characteristics of a facility creating a shared knowledge resource for information about it and forming a reliable basis for decisions during its life cycle, from earliest conception to demolition'.

### Business Case

The **Business Case** for a project is the rationale behind the initiation of a new building project. It may consist solely of a reasoned argument. It may contain supporting information, financial appraisals or other background information. It should also highlight initial considerations for the **Project Outcomes**. In summary, it is a combination of objective and subjective considerations. The **Business Case** might be prepared in relation to, for example, appraising a number of sites or in relation to assessing a refurbishment against a new build option.

### Change Control Procedures

Procedures for controlling changes to the design and construction following the sign-off of the Stage 2 Concept Design and the **Final Project Brief**.

### Common Standards

Publicly available standards frequently used to define project and design management processes in relation to the briefing, designing, constructing, maintaining, operating and use of a building.

### Communication Strategy

The strategy that sets out when the project team will meet, how they will communicate effectively and the protocols for issuing information between the various parties, both informally and at Information Exchanges.

## Construction Programme

The period in the **Project Programme** and the **Building Contract** for the construction of the project, commencing on the site mobilisation date and ending at **Practical Completion**.

## Construction Strategy

A strategy that considers specific aspects of the design that may affect the buildability or logistics of constructing a project, or may affect health and safety aspects. The **Construction Strategy** comprises items such as cranage, site access and accommodation locations, reviews of the supply chain and sources of materials, and specific buildability items, such as the choice of frame (steel or concrete) or the installation of larger items of plant. On a smaller project, the strategy may be restricted to the location of site cabins and storage, and the ability to transport materials up an existing staircase.

## Contractor's Proposals

Proposals presented by a contractor to the client in response to a tender that includes the **Employer's Requirements**. The **Contractor's Proposals** may match the **Employer's Requirements**, although certain aspects may be varied based on value engineered solutions and additional information may be submitted to clarify what is included in the tender. The **Contractor's Proposals** form an integral component of the **Building Contract** documentation.

## Contractual Tree

A diagram that clarifies the contractual relationship between the client and the parties undertaking the roles required on a project.

## Cost Information

All of the project costs, including the cost estimate and life cycle costs where required.

## Design Programme

A programme setting out the strategic dates in relation to the design process. It is aligned with the **Project Programme** but is strategic in its nature, due to the iterative nature of the design process, particularly in the early stages.

## Design Queries

Queries relating to the design arising from the site, typically managed using a contractor's in-house request for information (RFI) or technical query (TQ) process.

## Design Responsibility Matrix

A matrix that sets out who is responsible for designing each aspect of the project and when. This document sets out the extent of any performance specified design. The **Design Responsibility Matrix** is created at a strategic level at Stage 1 and fine tuned in response to the Concept Design at the end of Stage 2 in order to ensure that there are no design responsibility ambiguities at Stages 3, 4 and 5.

## Employer's Requirements

Proposals prepared by design team members. The level of detail will depend on the stage at which the tender is issued to the contractor. The **Employer's Requirements** may comprise a mixture of prescriptive elements and descriptive elements to allow the contractor a degree of flexibility in determining the **Contractor's Proposals**.

### Feasibility Studies

Studies undertaken on a given site to test the feasibility of the **Initial Project Brief** on a specific site or in a specific context and to consider how site-wide issues will be addressed.

### Feedback

**Feedback** from the project team, including the end users, following completion of a building.

### Final Project Brief

The **Initial Project Brief** amended so that it is aligned with the Concept Design and any briefing decisions made during Stage 2. (Both the Concept Design and **Initial Project Brief** are Information Exchanges at the end of Stage 2.)

### Handover Strategy

The strategy for handing over a building, including the requirements for phased handovers, commissioning, training of staff or other factors crucial to the successful occupation of a building. On some projects, the Building Services Research and Information Association (BSRIA) Soft Landings process is used as the basis for formulating the strategy and undertaking a **Post-occupancy Evaluation** (www.bsria. co.uk/services/design/soft-landings/).

### Health and Safety Strategy

The strategy covering all aspects of health and safety on the project, outlining legislative requirements as well as other project initiatives, including the **Maintenance and Operational Strategy**.

### Information Exchange

The formal issue of information for review and sign-off by the client at key stages of the project. The project team may also have additional formal **Information Exchanges** as well as the many informal exchanges that occur during the iterative design process.

### Initial Project Brief

The brief prepared following discussions with the client to ascertain the **Project Objectives**, the client's **Business Case** and, in certain instances, in response to site **Feasibility Studies**.

### Maintenance and Operational Strategy

The strategy for the maintenance and operation of a building, including details of any specific plant required to replace components.

### Post-occupancy Evaluation

Evaluation undertaken post occupancy to determine whether the **Project Outcomes**, both subjective and objective, set out in the **Final Project Brief** have been achieved.

### Practical Completion

**Practical Completion** is a contractual term used in the **Building Contract** to signify the date on which a project is handed over to the client. The date triggers a number of contractual mechanisms.

### Project Budget

The client's budget for the project, which may include the construction cost as well as the cost of certain items required post completion and during the project's operational use.

### Project Execution Plan

The **Project Execution Plan** is produced in collaboration between the project lead and lead designer, with contributions from other designers and members of the project team. The **Project Execution Plan** sets out the processes and protocols to be used to develop the design. It is sometimes referred to as a project quality plan.

## Project Information

Information, including models, documents, specifications, schedules and spreadsheets, issued between parties during each stage and in formal Information Exchanges at the end of each stage.

## Project Objectives

The client's key objectives as set out in the **Initial Project Brief**. The document includes, where appropriate, the employer's **Business Case**, **Sustainability Aspirations** or other aspects that may influence the preparation of the brief and, in turn, the Concept Design stage. For example, **Feasibility Studies** may be required in order to test the **Initial Project Brief** against a given site, allowing certain high-level briefing issues to be considered before design work commences in earnest.

## Project Outcomes

The desired outcomes for the project (for example, in the case of a hospital this might be a reduction in recovery times). The outcomes may include operational aspects and a mixture of subjective and objective criteria.

## Project Performance

The performance of the project, determined using **Feedback**, including about the performance of the project team and the performance of the building against the desired **Project Outcomes**.

## Project Programme

The overall period for the briefing, design, construction and post-completion activities of a project.

## Project Roles Table

A table that sets out the roles required on a project as well as defining the stages during which those roles are required and the parties responsible for carrying out the roles.

## Project Strategies

The strategies developed in parallel with the Concept Design to support the design and, in certain instances, to respond to the **Final Project Brief** as it is concluded. These strategies typically include:

I  acoustic strategy
I  fire engineering strategy
I  **Maintenance and Operational Strategy**
I  **Sustainability Strategy**
I  building control strategy
I  **Technology Strategy**.

These strategies are usually prepared in outline at Stage 2 and in detail at Stage 3, with the recommendations absorbed into the Stage 4 outputs and Information Exchanges.

The strategies are not typically used for construction purposes because they may contain recommendations or information that contradict the drawn information. The intention is that they should be transferred into the various models or drawn information.

## Quality Objectives

The objectives that set out the quality aspects of a project. The objectives may comprise both subjective and objective aspects, although subjective aspects may be subject to a design quality indicator (DQI) benchmark review during the **Feedback** period.

## Research and Development

Project-specific research and development responding to the **Initial Project Brief** or in response to the Concept Design as it is developed.

## Risk Assessment

The **Risk Assessment** considers the various design and other risks on a project and how each risk will be managed and the party responsible for managing each risk.

## Schedule of Services

A list of specific services and tasks to be undertaken by a party involved in the project which is incorporated into their professional services contract.

## Site Information

Specific **Project Information** in the form of specialist surveys or reports relating to the project- or site-specific context.

## Strategic Brief

The brief prepared to enable the Strategic Definition of the project. Strategic considerations might include considering different sites, whether to extend, refurbish or build new and the key **Project Outcomes** as well as initial considerations for the **Project Programme** and assembling the project team.

## Sustainability Aspirations

The client's aspirations for sustainability, which may include additional objectives, measures or specific levels of performance in relation to international standards, as well as details of specific demands in relation to operational or facilities management issues.

The **Sustainability Strategy** will be prepared in response to the **Sustainability Aspirations** and will include specific additional items, such as an energy plan and ecology plan and the design life of the building, as appropriate.

## Sustainability Strategy

The strategy for delivering the **Sustainability Aspirations**.

## Technology Strategy

The strategy established at the outset of a project that sets out technologies, including Building Information Modelling (BIM) and any supporting processes, and the specific software packages that each member of the project team will use. Any interoperability issues can then be addressed before the design phases commence.

This strategy also considers how information is to be communicated (by email, file transfer protocol (FTP) site or using a managed third party common data environment) as well as the file formats in which information will be provided. The **Project Execution Plan** records agreements made.

## Work in Progress

**Work in Progress** is ongoing design work that is issued between designers to facilitate the iterative coordination of each designer's output. Work issued as **Work in Progress** is signed off by the internal design processes of each designer and is checked and coordinated by the lead designer.

# Index